INSTRUCTOR'S SOLUTIONS MANUAL

STEVEN J. LEON
University of Massachusetts at Dartmouth

7E

LINEAR ALGEBRA
With Applications

Upper Saddle River, NJ 07458

Editor-in-Chief: Sally Yagan
Supplement Editor: Jennifer Urban
Executive Managing Editor: Kathleen Schiaparelli
Assistant Managing Editor: Karen Bosch
Production Editor: Jenelle J. Woodrup
Supplement Cover Manager: Paul Gourhan
Supplement Cover Designer: Christopher Kossa
Manufacturing Buyer: Ilene Kahn
Manufacturing Manager: Alexis Heydt-Long

Pearson Prentice Hall
Pearson Education, Inc.
Upper Saddle River, NJ 07458

Printed in the United States of America

10 9 8 7 6 5 4 3 2 1

ISBN 0-13-193625-5

Pearson Education Ltd., *London*
Pearson Education Australia Pty. Ltd., *Sydney*
Pearson Education Singapore, Pte. Ltd.
Pearson Education North Asia Ltd., *Hong Kong*
Pearson Education Canada, Inc., *Toronto*
Pearson Educación de Mexico, S.A. de C.V.
Pearson Education—Japan, *Tokyo*
Pearson Education Malaysia, Pte. Ltd.

TABLE OF CONTENTS

PREFACE

This solutions manual is designed to accompany the seventh edition of *Linear Algebra with Applications* by Steven J. Leon. The answers in this manual supplement those given in the answer key of the textbook. In addition this manual contains the complete solutions to all of the nonroutine exercises in the book.

At the end of each chapter of the textbook there are two chapter tests (A and B) and a section of computer exercises to be solved using MATLAB. The questions in each Chapter Test A are to be answered as either *true* or *false*. Although the true-false answers are given in the Answer Section of the textbook, students are required to explain or prove their answers. This manual includes explanations, proofs, and counterexamples for all Chapter Test A questions. The chapter tests labelled B contain workout problems. The answers to these problems are not given in the Answers to Selected Exercises Section of the textbook, however, they are provided in this manual. Complete solutions are given for all of the nonroutine Chapter Test B exercises.

In the MATLAB exercises most of the computations are straightforward. Consequently they have not been included in this solutions manual. On the other hand, the text also includes questions related to the computations. The purpose of the questions is to emphasize the significance of the computations. The solutions manual does provide the answers to most of these questions. There are some questions for which it is not possible to provide a single answer. For example, aome exercises involve randomly generated matrices. In these cases the answers may depend on the particular random matrices that were generated.

Steven J. Leon
sleon@umassd.edu

CHAPTER 1

SECTION 1

2. (d) $\begin{pmatrix} 1 & 1 & 1 & 1 & 1 \\ 0 & 2 & 1 & -2 & 1 \\ 0 & 0 & 4 & 1 & -2 \\ 0 & 0 & 0 & 1 & -3 \\ 0 & 0 & 0 & 0 & 2 \end{pmatrix}$

5. (a) $3x_1 + 2x_2 = 8$
$x_1 + 5x_2 = 7$

(b) $5x_1 - 2x_2 + \ x_3 = 3$
$2x_1 + 3x_2 - 4x_3 = 0$

(c) $2x_1 + \ x_2 + 4x_3 = -1$
$4x_1 - 2x_2 + 3x_3 = \ 4$
$5x_1 + 2x_2 + 6x_2 = -1$

(d) $4x_1 - 3x_2 + \ x_3 + 2x_4 = 4$
$3x_1 + \ x_2 - 5x_3 + 6x_4 = 5$
$x_1 + \ x_2 + 2x_3 + 4x_4 = 8$
$5x_1 + \ x_2 + 3x_3 - 2x_4 = 7$

9. Given the system

$$\begin{aligned} -m_1x_1 + x_2 &= b_1 \\ -m_2x_1 + x_2 &= b_2 \end{aligned}$$

one can eliminate the variable x_2 by subtracting the first row from the second. One then obtains the equivalent system

$$\begin{aligned} -m_1x_1 + x_2 &= b_1 \\ (m_1 - m_2)x_1 &= b_2 - b_1 \end{aligned}$$

(a) If $m_1 \neq m_2$, then one can solve the second equation for x_1

$$x_1 = \frac{b_2 - b_1}{m_1 - m_2}$$

One can then plug this value of x_1 into the first equation and solve for x_2. Thus, if $m_1 \neq m_2$, there will be a unique ordered pair (x_1, x_2) that satisfies the two equations.

(b) If $m_1 = m_2$, then the x_1 term drops out in the second equation

$$0 = b_2 - b_1$$

This is possible if and only if $b_1 = b_2$.

(c) If $m_1 \neq m_2$, then the two equations represent lines in the plane with different slopes. Two nonparallel lines intersect in a point. That point will be the unique solution to the system. If $m_1 = m_2$ and $b_1 = b_2$, then both equations represent the same line and consequently every point on that line will satisfy both equations. If $m_1 = m_2$ and $b_1 \neq b_2$, then the equations represent parallel lines. Since parallel lines do not intersect, there is no point on both lines and hence no solution to the system.

10. The system must be consistent since $(0, 0)$ is a solution.

11. A linear equation in 3 unknowns represents a plane in three space. The solution set to a 3×3 linear system would be the set of all points that lie on all three planes. If the planes are parallel or one plane is parallel to the line of intersection of the other two, then the solution set will be empty. The three equations could represent the same plane or the three planes could all intersect in a line. In either case the solution set will contain infinitely many points. If the three planes intersect in a point then the solution set will contain only that point.

SECTION 2

2. (b) The system is consistent with a unique solution $(4, -1)$.

4. (b) x_1 and x_3 are lead variables and x_2 is a free variable.
(d) x_1 and x_3 are lead variables and x_2 and x_4 are free variables.
(f) x_2 and x_3 are lead variables and x_1 is a free variable.

5. (l) The solution is $(0, -1.5, -3.5)$.

6. (c) The solution set consists of all ordered triples of the form $(0, -\alpha, \alpha)$.

7. A homogeneous linear equation in 3 unknowns corresponds to a plane that passes through the origin in 3-space. Two such equations would correspond to two planes through the origin. If one equation is a multiple of the other, then both represent the same plane through the origin and every point on that plane will be a solution to the system. If one equation is not a multiple of the other, then we have two distinct planes that intersect in a line through the origin. Every point on the line of intersection will be a solution to the linear system. So in either case the system must have infinitely many solutions.

In the case of a nonhomogeneous 2×3 linear system, the equations correspond to planes that do not both pass through the origin. If one equation is a multiple of the other, then both represent the same plane and there are infinitely many solutions. If the equations represent planes that are parallel, then they do not intersect and hence the system will not have any solutions. If the equations represent distinct planes that are not parallel, then they must intersect in a line and hence there will be infinitely many solutions. So the only possibilities for a nonhomogeneous 2×3 linear system are 0 or infinitely many solutions.

9. (a) Since the system is homogeneous it must be consistent.

14. At each intersection the number of vehicles entering must equal the number of vehicles leaving in order for the traffic to flow. This condition leads to the following system of equations

$$\begin{aligned} x_1 + a_1 &= x_2 + b_1 \\ x_2 + a_2 &= x_3 + b_2 \\ x_3 + a_3 &= x_4 + b_3 \\ x_4 + a_4 &= x_1 + b_4 \end{aligned}$$

If we add all four equations we get

$$x_1 + x_2 + x_3 + x_4 + a_1 + a_2 + a_3 + a_4 = x_1 + x_2 + x_3 + x_4 + b_1 + b_2 + b_3 + b_4$$

and hence

$$a_1 + a_2 + a_3 + a_4 = b_1 + b_2 + b_3 + b_4$$

15. If (c_1, c_2) is a solution, then

$$\begin{aligned} a_{11}c_1 + a_{12}c_2 &= 0 \\ a_{21}c_1 + a_{22}c_2 &= 0 \end{aligned}$$

Multiplying both equations through by α, one obtains

$$\begin{aligned} a_{11}(\alpha c_1) + a_{12}(\alpha c_2) &= \alpha \cdot 0 = 0 \\ a_{21}(\alpha c_1) + a_{22}(\alpha c_2) &= \alpha \cdot 0 = 0 \end{aligned}$$

Thus $(\alpha c_1, \alpha c_2)$ is also a solution.

16. (a) If $x_4 = 0$ then x_1, x_2, and x_3 will all be 0. Thus if no glucose is produced then there is no reaction. $(0, 0, 0, 0)$ is the trivial solution in the sense that if there are no molecules of carbon dioxide and water, then there will be no reaction.

(b) If we choose another value of x_4, say $x_4 = 2$, then we end up with solution $x_1 = 12$, $x_2 = 12$, $x_3 = 12$, $x_4 = 2$. Note the ratios are still 6:6:6:1.

SECTION 3

1. (e) $\begin{pmatrix} 8 & -15 & 11 \\ 0 & -4 & -3 \\ -1 & -6 & 6 \end{pmatrix}$

(g) $\begin{pmatrix} 5 & -10 & 15 \\ 5 & -1 & 4 \\ 8 & -9 & 6 \end{pmatrix}$

2. (d) $\begin{pmatrix} 36 & 10 & 56 \\ 10 & 3 & 16 \end{pmatrix}$

5. (a) $5A = \begin{pmatrix} 15 & 20 \\ 5 & 5 \\ 10 & 35 \end{pmatrix}$

$$2A + 3A = \begin{pmatrix} 6 & 8 \\ 2 & 2 \\ 4 & 14 \end{pmatrix} + \begin{pmatrix} 9 & 12 \\ 3 & 3 \\ 6 & 21 \end{pmatrix} = \begin{pmatrix} 15 & 20 \\ 5 & 5 \\ 10 & 35 \end{pmatrix}$$

(b) $6A = \begin{pmatrix} 18 & 24 \\ 6 & 6 \\ 12 & 42 \end{pmatrix}$

$$3(2A) = 3\begin{pmatrix} 6 & 8 \\ 2 & 2 \\ 4 & 14 \end{pmatrix} = \begin{pmatrix} 18 & 24 \\ 6 & 6 \\ 12 & 42 \end{pmatrix}$$

(c) $A^T = \begin{pmatrix} 3 & 1 & 2 \\ 4 & 1 & 7 \end{pmatrix}$

$$(A^T)^T = \begin{pmatrix} 3 & 1 & 2 \\ 4 & 1 & 7 \end{pmatrix}^T = \begin{pmatrix} 3 & 4 \\ 1 & 1 \\ 2 & 7 \end{pmatrix} = A$$

6. (a) $A + B = \begin{pmatrix} 5 & 4 & 6 \\ 0 & 5 & 1 \end{pmatrix} = B + A$

(b) $3(A + B) = 3\begin{pmatrix} 5 & 4 & 6 \\ 0 & 5 & 1 \end{pmatrix} = \begin{pmatrix} 15 & 12 & 18 \\ 0 & 15 & 3 \end{pmatrix}$

$$3A + 3B = \begin{pmatrix} 12 & 3 & 18 \\ 6 & 9 & 15 \end{pmatrix} + \begin{pmatrix} 3 & 9 & 0 \\ -6 & 6 & -12 \end{pmatrix} = \begin{pmatrix} 15 & 12 & 18 \\ 0 & 15 & 3 \end{pmatrix}$$

(c) $(A + B)^T = \begin{pmatrix} 5 & 4 & 6 \\ 0 & 5 & 1 \end{pmatrix}^T = \begin{pmatrix} 5 & 0 \\ 4 & 5 \\ 6 & 1 \end{pmatrix}$

$$A^T + B^T = \begin{pmatrix} 4 & 2 \\ 1 & 3 \\ 6 & 5 \end{pmatrix} + \begin{pmatrix} 1 & -2 \\ 3 & 2 \\ 0 & -4 \end{pmatrix} = \begin{pmatrix} 5 & 0 \\ 4 & 5 \\ 6 & 1 \end{pmatrix}$$

7. (a) $3(AB) = 3\begin{pmatrix} 5 & 14 \\ 15 & 42 \\ 0 & 16 \end{pmatrix} = \begin{pmatrix} 15 & 42 \\ 45 & 126 \\ 0 & 48 \end{pmatrix}$

$$(3A)B = \begin{pmatrix} 6 & 3 \\ 18 & 9 \\ -6 & 12 \end{pmatrix}\begin{pmatrix} 2 & 4 \\ 1 & 6 \end{pmatrix} = \begin{pmatrix} 15 & 42 \\ 45 & 126 \\ 0 & 48 \end{pmatrix}$$

$$A(3B) = \begin{pmatrix} 2 & 1 \\ 6 & 3 \\ -2 & 4 \end{pmatrix} \begin{pmatrix} 6 & 12 \\ 3 & 18 \end{pmatrix} = \begin{pmatrix} 15 & 42 \\ 45 & 126 \\ 0 & 48 \end{pmatrix}$$

(b) $$(AB)^T = \begin{pmatrix} 5 & 14 \\ 15 & 42 \\ 0 & 16 \end{pmatrix}^T = \begin{pmatrix} 5 & 15 & 0 \\ 14 & 42 & 16 \end{pmatrix}$$

$$B^T A^T = \begin{pmatrix} 2 & 1 \\ 4 & 6 \end{pmatrix} \begin{pmatrix} 2 & 6 & -2 \\ 1 & 3 & 4 \end{pmatrix} = \begin{pmatrix} 5 & 15 & 0 \\ 14 & 42 & 16 \end{pmatrix}$$

8. (a) $$(A+B)+C = \begin{pmatrix} 0 & 5 \\ 1 & 7 \end{pmatrix} + \begin{pmatrix} 3 & 1 \\ 2 & 1 \end{pmatrix} = \begin{pmatrix} 3 & 6 \\ 3 & 8 \end{pmatrix}$$

$$A+(B+C) = \begin{pmatrix} 2 & 4 \\ 1 & 3 \end{pmatrix} + \begin{pmatrix} 1 & 2 \\ 2 & 5 \end{pmatrix} = \begin{pmatrix} 3 & 6 \\ 3 & 8 \end{pmatrix}$$

(b) $$(AB)C = \begin{pmatrix} -4 & 18 \\ -2 & 13 \end{pmatrix} \begin{pmatrix} 3 & 1 \\ 2 & 1 \end{pmatrix} = \begin{pmatrix} 24 & 14 \\ 20 & 11 \end{pmatrix}$$

$$A(BC) = \begin{pmatrix} 2 & 4 \\ 1 & 3 \end{pmatrix} \begin{pmatrix} -4 & -1 \\ 8 & 4 \end{pmatrix} = \begin{pmatrix} 24 & 14 \\ 20 & 11 \end{pmatrix}$$

(c) $$A(B+C) = \begin{pmatrix} 2 & 4 \\ 1 & 3 \end{pmatrix} \begin{pmatrix} 1 & 2 \\ 2 & 5 \end{pmatrix} = \begin{pmatrix} 10 & 24 \\ 7 & 17 \end{pmatrix}$$

$$AB + AC = \begin{pmatrix} -4 & 18 \\ -2 & 13 \end{pmatrix} + \begin{pmatrix} 14 & 6 \\ 9 & 4 \end{pmatrix} = \begin{pmatrix} 10 & 24 \\ 7 & 17 \end{pmatrix}$$

(d) $$(A+B)C = \begin{pmatrix} 0 & 5 \\ 1 & 7 \end{pmatrix} \begin{pmatrix} 3 & 1 \\ 2 & 1 \end{pmatrix} = \begin{pmatrix} 10 & 5 \\ 17 & 8 \end{pmatrix}$$

$$AC + BC = \begin{pmatrix} 14 & 6 \\ 9 & 4 \end{pmatrix} + \begin{pmatrix} -4 & -1 \\ 8 & 4 \end{pmatrix} = \begin{pmatrix} 10 & 5 \\ 17 & 8 \end{pmatrix}$$

9. Let

$$D = (AB)C = \begin{pmatrix} a_{11}b_{11} + a_{12}b_{21} & a_{11}b_{12} + a_{12}b_{22} \\ a_{21}b_{11} + a_{22}b_{21} & a_{21}b_{12} + a_{22}b_{22} \end{pmatrix} \begin{pmatrix} c_{11} & c_{12} \\ c_{21} & c_{22} \end{pmatrix}$$

It follows that

$$\begin{aligned} d_{11} &= (a_{11}b_{11} + a_{12}b_{21})c_{11} + (a_{11}b_{12} + a_{12}b_{22})c_{21} \\ &= a_{11}b_{11}c_{11} + a_{12}b_{21}c_{11} + a_{11}b_{12}c_{21} + a_{12}b_{22}c_{21} \\ d_{12} &= (a_{11}b_{11} + a_{12}b_{21})c_{12} + (a_{11}b_{12} + a_{12}b_{22})c_{22} \\ &= a_{11}b_{11}c_{12} + a_{12}b_{21}c_{12} + a_{11}b_{12}c_{22} + a_{12}b_{22}c_{22} \\ d_{21} &= (a_{21}b_{11} + a_{22}b_{21})c_{11} + (a_{21}b_{12} + a_{22}b_{22})c_{21} \\ &= a_{21}b_{11}c_{11} + a_{22}b_{21}c_{11} + a_{21}b_{12}c_{21} + a_{22}b_{22}c_{21} \\ d_{22} &= (a_{21}b_{11} + a_{22}b_{21})c_{12} + (a_{21}b_{12} + a_{22}b_{22})c_{22} \\ &= a_{21}b_{11}c_{12} + a_{22}b_{21}c_{12} + a_{21}b_{12}c_{22} + a_{22}b_{22}c_{22} \end{aligned}$$

If we set

$$E = A(BC) = \begin{pmatrix} a_{11} & a_{12} \\ a_{21} & a_{22} \end{pmatrix} \begin{pmatrix} b_{11}c_{11} + b_{12}c_{21} & b_{11}c_{12} + b_{12}c_{22} \\ b_{21}c_{11} + b_{22}c_{21} & b_{21}c_{12} + b_{22}c_{22} \end{pmatrix}$$

then it follows that

$$\begin{aligned}
e_{11} &= a_{11}(b_{11}c_{11} + b_{12}c_{21}) + a_{12}(b_{21}c_{11} + b_{22}c_{21}) \\
&= a_{11}b_{11}c_{11} + a_{11}b_{12}c_{21} + a_{12}b_{21}c_{11} + a_{12}b_{22}c_{21} \\
e_{12} &= a_{11}(b_{11}c_{12} + b_{12}c_{22}) + a_{12}(b_{21}c_{12} + b_{22}c_{22}) \\
&= a_{11}b_{11}c_{12} + a_{11}b_{12}c_{22} + a_{12}b_{21}c_{12} + a_{12}b_{22}c_{22} \\
e_{21} &= a_{21}(b_{11}c_{11} + b_{12}c_{21}) + a_{22}(b_{21}c_{11} + b_{22}c_{21}) \\
&= a_{21}b_{11}c_{11} + a_{21}b_{12}c_{21} + a_{22}b_{21}c_{11} + a_{22}b_{22}c_{21} \\
e_{22} &= a_{21}(b_{11}c_{12} + b_{12}c_{22}) + a_{22}(b_{21}c_{12} + b_{22}c_{22}) \\
&= a_{21}b_{11}c_{12} + a_{21}b_{12}c_{22} + a_{22}b_{21}c_{12} + a_{22}b_{22}c_{22}
\end{aligned}$$

Thus

$$d_{11} = e_{11} \qquad d_{12} = e_{12} \qquad d_{21} = e_{21} \qquad d_{22} = e_{22}$$

and hence

$$(AB)C = D = E = A(BC)$$

12.

$$A^2 = \begin{pmatrix} 0 & 0 & 1 & 0 \\ 0 & 0 & 0 & 1 \\ 0 & 0 & 0 & 0 \\ 0 & 0 & 0 & 0 \end{pmatrix} \qquad A^3 = \begin{pmatrix} 0 & 0 & 0 & 1 \\ 0 & 0 & 0 & 0 \\ 0 & 0 & 0 & 0 \\ 0 & 0 & 0 & 0 \end{pmatrix}$$

and $A^4 = O$. If $n > 4$, then

$$A^n = A^{n-4}A^4 = A^{n-4}O = O$$

13. (b) $\mathbf{x} = (2, 1)^T$ is a solution since $\mathbf{b} = 2\mathbf{a}_1 + \mathbf{a}_2$. There are no other solutions since the echelon form of A is strictly triangular.
(c) The solution to $A\mathbf{x} = \mathbf{c}$ is $\mathbf{x} = (-\frac{5}{2}, -\frac{1}{4})^T$. Therefore $\mathbf{c} = -\frac{5}{2}\mathbf{a}_1 - \frac{1}{4}\mathbf{a}_2$.

15. If $d = a_{11}a_{22} - a_{21}a_{12} \neq 0$ then

$$\frac{1}{d}\begin{pmatrix} a_{22} & -a_{12} \\ -a_{21} & a_{11} \end{pmatrix}\begin{pmatrix} a_{11} & a_{12} \\ a_{21} & a_{22} \end{pmatrix} = \begin{pmatrix} \frac{a_{11}a_{22} - a_{12}a_{21}}{d} & 0 \\ 0 & \frac{a_{11}a_{22} - a_{12}a_{21}}{d} \end{pmatrix} = I$$

$$\begin{pmatrix} a_{11} & a_{12} \\ a_{21} & a_{22} \end{pmatrix}\left[\frac{1}{d}\begin{pmatrix} a_{22} & -a_{12} \\ -a_{21} & a_{11} \end{pmatrix}\right] = \begin{pmatrix} \frac{a_{11}a_{22} - a_{12}a_{21}}{d} & 0 \\ 0 & \frac{a_{11}a_{22} - a_{12}a_{21}}{d} \end{pmatrix} = I$$

Therefore

$$\frac{1}{d}\begin{pmatrix} a_{22} & -a_{12} \\ -a_{21} & a_{11} \end{pmatrix} = A^{-1}$$

16. Since

$$A^{-1}A = AA^{-1} = I$$

it follows from the definition that A^{-1} is nonsingular and its inverse is A.

17. Since

$$\begin{aligned} A^T(A^{-1})^T &= (A^{-1}A)^T = I \\ (A^{-1})^T A^T &= (AA^{-1})^T = I \end{aligned}$$

it follows that

$$(A^{-1})^T = (A^T)^{-1}$$

18. If $A\mathbf{x} = A\mathbf{y}$ and $\mathbf{x} \neq \mathbf{y}$, then A must be singular, for if A were nonsingular then we could multiply by A^{-1} and get

$$\begin{aligned} A^{-1}A\mathbf{x} &= A^{-1}A\mathbf{y} \\ \mathbf{x} &= \mathbf{y} \end{aligned}$$

19. For $m = 1$,

$$(A^1)^{-1} = A^{-1} = (A^{-1})^1$$

Assume the result holds in the case $m = k$, that is,

$$(A^k)^{-1} = (A^{-1})^k$$

It follows that

$$(A^{-1})^{k+1}A^{k+1} = A^{-1}(A^{-1})^k A^k A = A^{-1}A = I$$

and

$$A^{k+1}(A^{-1})^{k+1} = AA^k(A^{-1})^k A^{-1} = AA^{-1} = I$$

Therefore

$$(A^{-1})^{k+1} = (A^{k+1})^{-1}$$

and the result follows by mathematical induction.

20. **(a)** $(A+B)^2 = (A+B)(A+B) = (A+B)A+(A+B)B = A^2+BA+AB+B^2$
In the case of real numbers $ab + ba = 2ab$, however, with matrices $AB + BA$ is generally not equal to $2AB$.

(b)

$$\begin{aligned} (A+B)(A-B) &= (A+B)(A-B) \\ &= (A+B)A - (A+B)B \\ &= A^2 + BA - AB - B^2 \end{aligned}$$

In the case of real numbers $ab-ba = 0$, however, with matrices $AB-BA$ is generally not equal to O.

21. If we replace a by A and b by the identity matrix, I, then both rules will work, since

$$(A+I)^2 = A^2 + IA + AI + B^2 = A^2 + AI + AI + B^2 = A^2 + 2AI + B^2$$

and

$$(A+I)(A-I) = A^2 + IA - AI - I^2 = A^2 + A - A - I^2 = A^2 - I^2$$

22. There are many possible choices for A and B. For example, one could choose

$$A = \begin{pmatrix} 0 & 1 \\ 0 & 0 \end{pmatrix} \quad \text{and} \quad B = \begin{pmatrix} 1 & 1 \\ 0 & 0 \end{pmatrix}$$

More generally if

$$A = \begin{pmatrix} a & b \\ ca & cb \end{pmatrix} \qquad B = \begin{pmatrix} db & eb \\ -da & -ea \end{pmatrix}$$

then $AB = O$ for any choice of the scalars a, b, c, d, e.

23. To construct nonzero matrices A, B, C with the desired properties, first find nonzero matrices C and D such that $DC = O$ (see Exercise 22). Next, for any nonzero matrix A, set $B = A + D$. It follows that

$$BC = (A + D)C = AC + DC = AC + O = AC$$

24. A 2×2 symmetric matrix is one of the form

$$A = \begin{pmatrix} a & b \\ b & c \end{pmatrix}$$

Thus

$$A^2 = \begin{pmatrix} a^2 + b^2 & ab + bc \\ ab + bc & b^2 + c^2 \end{pmatrix}$$

If $A^2 = O$, then its diagonal entries must be 0.

$$a^2 + b^2 = 0 \quad \text{and} \quad b^2 + c^2 = 0$$

Thus $a = b = c = 0$ and hence $A = O$.

25. For most pairs of symmetric matrices A and B the product AB will not be symmetric. For example

$$\begin{pmatrix} 1 & 1 \\ 1 & 2 \end{pmatrix} \begin{pmatrix} 1 & 2 \\ 2 & 1 \end{pmatrix} = \begin{pmatrix} 3 & 3 \\ 5 & 4 \end{pmatrix}$$

See Exercise 27 for a characterization of the conditions under which the product will be symmetric.

26. (a) A^T is an $n \times m$ matrix. Since A^T has m columns and A has m rows, the multiplication A^TA is possible. The multiplication AA^T is possible since A has n columns and A^T has n rows.

(b) $(A^TA)^T = A^T(A^T)^T = A^TA$
$(AA^T)^T = (A^T)^TA^T = AA^T$

27. Let A and B be symmetric $n \times n$ matrices. If $(AB)^T = AB$ then

$$BA = B^TA^T = (AB)^T = AB$$

Conversely if $BA = AB$ then

$$(AB)^T = B^TA^T = BA = AB$$

28. If A is skew-symmetric then $A^T = -A$. Since the (j, j) entry of A^T is a_{jj} and the (j, j) entry of $-A$ is $-a_{jj}$, it follows that is $a_{jj} = -a_{jj}$ for each j and hence the diagonal entries of A must all be 0.

29. (a)

$$\begin{aligned} B^T &= (A + A^T)^T = A^T + (A^T)^T = A^T + A = B \\ C^T &= (A - A^T)^T = A^T - (A^T)^T = A^T - A = -C \end{aligned}$$

(b) $A = \frac{1}{2}(A + A^T) + \frac{1}{2}(A - A^T)$

31. The search vector is $\mathbf{x} = (1, 0, 1, 0, 1, 0)^T$. The search result is given by the vector

$$\mathbf{y} = A^T\mathbf{x} = (1, 2, 2, 1, 1, 2, 1)^T$$

The ith entry of $\mathbf{y}$ is equal to the number of search words in the title of the ith book.

34. If $\alpha = a_{21}/a_{11}$, then

$$\begin{pmatrix} 1 & 0 \\ \alpha & 1 \end{pmatrix} \begin{pmatrix} a_{11} & a_{12} \\ 0 & b \end{pmatrix} = \begin{pmatrix} a_{11} & a_{12} \\ \alpha a_{11} & \alpha a_{12} + b \end{pmatrix} = \begin{pmatrix} a_{11} & a_{12} \\ a_{21} & \alpha a_{12} + b \end{pmatrix}$$

The product will equal A provided

$$\alpha a_{12} + b = a_{22}$$

Thus we must choose

$$b = a_{22} - \alpha a_{12} = a_{22} - \frac{a_{21}a_{12}}{a_{11}}$$

SECTION 4

2. (a) $\begin{pmatrix} 0 & 1 \\ 1 & 0 \end{pmatrix}$, type I

(b) The given matrix is not an elementary matrix. Its inverse is given by

$$\begin{pmatrix} \frac{1}{2} & 0 \\ 0 & \frac{1}{3} \end{pmatrix}$$

(c) $\begin{pmatrix} 1 & 0 & 0 \\ 0 & 1 & 0 \\ -5 & 0 & 1 \end{pmatrix}$, type III

(d) $\begin{pmatrix} 1 & 0 & 0 \\ 0 & 1/5 & 0 \\ 0 & 0 & 1 \end{pmatrix}$, type II

5. (c) Since

$$C = FB = FEA$$

where F and E are elementary matrices, it follows that C is row equivalent to A.

6. (b) $E_1^{-1} = \begin{pmatrix} 1 & 0 & 0 \\ 3 & 1 & 0 \\ 0 & 0 & 1 \end{pmatrix}$, $E_2^{-1} = \begin{pmatrix} 1 & 0 & 0 \\ 0 & 1 & 0 \\ 2 & 0 & 1 \end{pmatrix}$, $E_3^{-1} = \begin{pmatrix} 1 & 0 & 0 \\ 0 & 1 & 0 \\ 0 & -1 & 1 \end{pmatrix}$

The product $L = E_1^{-1}E_2^{-1}E_3^{-1}$ is lower triangular.

$$L = \begin{pmatrix} 1 & 0 & 0 \\ 3 & 1 & 0 \\ 2 & -1 & 1 \end{pmatrix}$$

7. A can be reduced to the identity matrix using three row operations

$$\begin{pmatrix} 2 & 1 \\ 6 & 4 \end{pmatrix} \to \begin{pmatrix} 2 & 1 \\ 0 & 1 \end{pmatrix} \to \begin{pmatrix} 2 & 0 \\ 0 & 1 \end{pmatrix} \to \begin{pmatrix} 1 & 0 \\ 0 & 1 \end{pmatrix}$$

The elementary matrices corresponding to the three row operations are

$$E_1 = \begin{pmatrix} 1 & 0 \\ -3 & 1 \end{pmatrix}, \quad E_2 = \begin{pmatrix} 1 & -1 \\ 0 & 1 \end{pmatrix}, \quad E_3 = \begin{pmatrix} \frac{1}{2} & 0 \\ 0 & 1 \end{pmatrix}$$

So

$$E_3E_2E_1A = I$$

and hence

$$A = E_1^{-1}E_3^{-1}E_3^{-1} = \begin{pmatrix} 1 & 0 \\ 3 & 1 \end{pmatrix} \begin{pmatrix} 1 & 1 \\ 0 & 1 \end{pmatrix} \begin{pmatrix} 2 & 0 \\ 0 & 1 \end{pmatrix}$$

and $A^{-1} = E_3E_2E_1$.

8. (b) $\begin{pmatrix} 1 & 0 \\ -1 & 1 \end{pmatrix} \begin{pmatrix} 2 & 4 \\ 0 & 5 \end{pmatrix}$

(d) $\begin{pmatrix} 1 & 0 & 0 \\ -2 & 1 & 0 \\ 3 & -2 & 1 \end{pmatrix} \begin{pmatrix} -2 & 1 & 2 \\ 0 & 3 & 2 \\ 0 & 0 & 2 \end{pmatrix}$

9. (a) $\begin{pmatrix} 1 & 0 & 1 \\ 3 & 3 & 4 \\ 2 & 2 & 3 \end{pmatrix} \begin{pmatrix} 1 & 2 & -3 \\ -1 & 1 & -1 \\ 0 & -2 & 3 \end{pmatrix} = \begin{pmatrix} 1 & 0 & 0 \\ 0 & 1 & 0 \\ 0 & 0 & 1 \end{pmatrix}$

$$\begin{pmatrix} 1 & 2 & -3 \\ -1 & 1 & -1 \\ 0 & -2 & -3 \end{pmatrix} \begin{pmatrix} 1 & 0 & 1 \\ 3 & 3 & 4 \\ 2 & 2 & 3 \end{pmatrix} = \begin{pmatrix} 1 & 0 & 0 \\ 0 & 1 & 0 \\ 0 & 0 & 1 \end{pmatrix}$$

10. (e) $\begin{pmatrix} 1 & -1 & 0 \\ 0 & 1 & -1 \\ 0 & 0 & 1 \end{pmatrix}$

12. (b) $XA + B = C$

$X = (C - B)A^{-1}$

$= \begin{pmatrix} 8 & -14 \\ -13 & 19 \end{pmatrix}$

(d) $XA + C = X$

$XA - XI = -C$

$X(A - I) = -C$

$X = -C(A - I)^{-1}$

$= \begin{pmatrix} 2 & -4 \\ -3 & 6 \end{pmatrix}$

13. (a) If E is an elementary matrix of type I or type II then E is symmetric. Thus $E^T = E$ is an elementary matrix of the same type. If E is the elementary matrix of type III formed by adding α times the ith row of the identity matrix to the jth row, then E^T is the elementary matrix of type III formed from the identity matrix by adding α times the jth row to the ith row.

(b) In general the product of two elementary matrices will not be an elementary matrix. Generally the product of two elementary matrices will be a matrix formed from the identity matrix by the performance of two row operations. For example, if

$$E_1 = \begin{pmatrix} 1 & 0 & 0 \\ 2 & 1 & 0 \\ 0 & 0 & 0 \end{pmatrix} \quad \text{and} \quad E_2 = \begin{pmatrix} 1 & 0 & 0 \\ 0 & 1 & 0 \\ 2 & 0 & 1 \end{pmatrix}$$

then E_1 and E_2 are elementary matrices, but

$$E_1E_2 = \begin{pmatrix} 1 & 0 & 0 \\ 2 & 1 & 0 \\ 2 & 0 & 1 \end{pmatrix}$$

is not an elementary matrix.

14. If $T = UR$, then

$$t_{ij} = \sum_{k=1}^{n} u_{ik}r_{kj}$$

Since U and R are upper triangular

$$\begin{aligned} u_{i1} &= u_{i2} = \cdots = u_{i,i-1} = 0 \\ r_{j+1,j} &= r_{j+2,j} = \cdots - r_{nj} = 0 \end{aligned}$$

If $i > j$, then

$$\begin{aligned} t_{ij} &= \sum_{k=1}^{j} u_{ik}r_{kj} + \sum_{k=j+1}^{n} u_{ik}r_{kj} \\ &= \sum_{k=1}^{j} 0\, r_{kj} + \sum_{k=j+1}^{n} u_{ik}0 \\ &= 0 \end{aligned}$$

Therefore T is upper triangular.

If $i = j$, then

$$\begin{aligned} t_{jj} = t_{ij} &= \sum_{k=1}^{i-1} u_{ik}r_{kj} + u_{jj}r_{jj} + \sum_{k=j+1}^{n} u_{ik}r_{kj} \\ &= \sum_{k=1}^{i-1} 0\, r_{kj} + u_{jj}r_{jj} + \sum_{k=j+1}^{n} u_{ik}0 \\ &= u_{jj}r_{jj} \end{aligned}$$

Therefore

$$t_{jj} = u_{jj} r_{jj} \qquad j = 1, \ldots, n$$

15. If we set $\mathbf{x} = (2, 1 - 4)^T$, then

$$A\mathbf{x} = 2\mathbf{a}_1 + 1\mathbf{a}_2 - 4\mathbf{a}_3 = \mathbf{0}$$

Thus $\mathbf{x}$ is a nonzero solution to the system $A\mathbf{x} = \mathbf{0}$. But if a homogeneous system has a nonzero solution, then it must have infinitely many solutions. In particular, if c is any scalar, then $c\mathbf{x}$ is also a solution to the system since

$$A(c\mathbf{x}) = cA\mathbf{x} = c\mathbf{0} = \mathbf{0}$$

Since $A\mathbf{x} = \mathbf{0}$ and $\mathbf{x} \neq \mathbf{0}$ it follows that the matrix A must be singular. (See Theorem 1.4.2)

16. If $\mathbf{a}_1 = 3\mathbf{a}_2 - 2\mathbf{a}_3$, then

$$\mathbf{a}_1 - 3\mathbf{a}_2 + 2\mathbf{a}_3 = \mathbf{0}$$

Therefore $\mathbf{x} = (1, -3, 2)^T$ is a nontrivial solution to $A\mathbf{x} = \mathbf{0}$. It follows form Theorem 1.4.2 that A must be singular.

17. If $\mathbf{x}_0 \neq \mathbf{0}$ and $A\mathbf{x}_0 = B\mathbf{x}_0$, then $C\mathbf{x}_0 = \mathbf{0}$ and it follows from Theorem 1.4.2 that C must be singular.

18. If B is singular, then it follows from Theorem 1.4.2 that there exists a nonzero vector $\mathbf{x}$ such that $B\mathbf{x} = \mathbf{0}$. If $C = AB$, then

$$C\mathbf{x} = AB\mathbf{x} = A\mathbf{0} = \mathbf{0}$$

Thus, by Theorem 1.4.2, C must also be singular.

19. (a) If U is upper triangular with nonzero diagonal entries, then using row operation II, U can be transformed into an upper triangular matrix with 1's on the diagonal. Row operation III can then be used to eliminate all of the entries above the diagonal. Thus U is row equivalent to I and hence is nonsingular.

(b) The same row operations that were used to reduce U to the identity matrix will transform I into U^{-1}. Row operation II applied to I will just change the values of the diagonal entries. When the row operation III steps referred to in part (a) are applied to a diagonal matrix, the entries above the diagonal are filled in. The resulting matrix, U^{-1}, will be upper triangular.

20. Since A is nonsingular it is row equivalent to I. Hence there exist elementary matrices $E_1, E_2, \ldots, E_k$ such that

$$E_k \cdots E_1 A = I$$

It follows that

$$A^{-1} = E_k \cdots E_1$$

and

$$E_k \cdots E_1 B = A^{-1} B = C$$

The same row operations that reduce A to I, will transform B to C. Therefore the reduced row echelon form of $(A \mid B)$ will be $(I \mid C)$.

21. (a) If the diagonal entries of D_1 are $\alpha_1, \alpha_2, \ldots, \alpha_n$ and the diagonal entries of D_2 are $\beta_1, \beta_2, \ldots, \beta_n$, then $D_1 D_2$ will be a diagonal matrix with diagonal entries $\alpha_1\beta_1, \alpha_2\beta_2, \ldots, \alpha_n\beta_n$ and $D_2 D_1$ will be a diagonal matrix with diagonal entries $\beta_1\alpha_1, \beta_2\alpha_2, \ldots, \beta_n\alpha_n$. Since the two have the same diagonal entries it follows that $D_1 D_2 = D_2 D_1$.

(b)

$$\begin{aligned} AB &= A(a_0 I + a_1 A + \cdots + a_k A^k) \\ &= a_0 A + a_1 A^2 + \cdots + a_k A^{k+1} \\ &= (a_0 I + a_1 A + \cdots + a_k A^k)A \\ &= BA \end{aligned}$$

22. If A is symmetric and nonsingular, then

$$(A^{-1})^T = (A^{-1})^T(AA^{-1}) = ((A^{-1})^T A^T)A^{-1} = A^{-1}$$

23. If A is row equivalent to B then there exist elementary matrices $E_1, E_2, \ldots, E_k$ such that

$$A = E_k E_{k-1} \cdots E_1 B$$

Each of the E_i's is invertible and E_i^{-1} is also an elementary matrix (Theorem 1.4.1). Thus

$$B = E_1^{-1} E_2^{-1} \cdots E_k^{-1} A$$

and hence B is row equivalent to A.

24. (a) If A is row equivalent to B, then there exist elementary matrices $E_1, E_2, \ldots, E_k$ such that

$$A = E_k E_{k-1} \cdots E_1 B$$

Since B is row equivalent to C, there exist elementary matrices $H_1, H_2, \ldots, H_j$ such that

$$B = H_j H_{j-1} \cdots H_1 C$$

Thus

$$A = E_k E_{k-1} \cdots E_1 H_j H_{j-1} \cdots H_1 C$$

and hence A is row equivalent to C.

(b) If A and B are nonsingular $n \times n$ matrices then A and B are row equivalent to I. Since A is row equivalent to I and I is row equivalent to B it follows from part (a) that A is row equivalent to B.

25. If U is any row echelon form of A then A can be reduced to U using row operations, so A is row equivalent to U. If B is row equivalent to A then it follows from the result in Exercise 24(a) that B is row equivalent to U.

26. If B is row equivalent to A, then there exist elementary matrices $E_1, E_2, \ldots, E_k$ such that

$$B = E_k E_{k-1} \cdots E_1 A$$

Let $M = E_k E_{k-1} \cdots E_1$. The matrix M is nonsingular since each of the E_i's is nonsingular.

Conversely suppose there exists a nonsingular matrix M such that $B = MA$. Since M is nonsingular it is row equivalent to I. Thus there exist elementary matrices $E_1, E_2, \ldots, E_k$ such that

$$M = E_k E_{k-1} \cdots E_1 I$$

It follows that

$$B = MA = E_k E_{k-1} \cdots E_1 A$$

Therefore B is row equivalent to A.

27. (a) The system $V\mathbf{c} = \mathbf{y}$ is given by

$$\begin{pmatrix} 1 & x_1 & x_1^2 & \cdots & x_1^n \\ 1 & x_2 & x_2^2 & \cdots & x_2^n \\ \vdots & & & & \\ 1 & x_{n+1} & x_{n+1}^2 & \cdots & x_{n+1}^n \end{pmatrix} \begin{pmatrix} c_1 \\ c_2 \\ \vdots \\ c_{n+1} \end{pmatrix} = \begin{pmatrix} y_1 \\ y_2 \\ \vdots \\ y_{n+1} \end{pmatrix}$$

Comparing the ith row of each side, we have

$$c_1 + c_2 x_i + \cdots + c_{n+1} x_i^n = y_i$$

Thus

$$p(x_i) = y_i \qquad i = 1, 2, \ldots, n+1$$

(b) If $x_1, x_2, \ldots, x_{n+1}$ are distinct and $V\mathbf{c} = \mathbf{0}$, then we can apply part (a) with $\mathbf{y} = \mathbf{0}$. Thus if $p(x) = c_1 + c_2 x + \cdots + c_{n+1} x^n$, then

$$p(x_i) = 0 \qquad i = 1, 2, \ldots, n+1$$

The polynomial $p(x)$ has $n+1$ roots. Since the degree of $p(x)$ is less than $n+1$, $p(x)$ must be the zero polynomial. Hence

$$c_1 = c_2 = \cdots = c_{n+1} = 0$$

Since the system $V\mathbf{c} = \mathbf{0}$ has only the trivial solution, the matrix V must be nonsingular.

SECTION 5

2. $B = A^T A = \begin{pmatrix} \mathbf{a}_1^T \\ \mathbf{a}_2^T \\ \vdots \\ \mathbf{a}_n^T \end{pmatrix} (\mathbf{a}_1, \mathbf{a}_2, \ldots, \mathbf{a}_n) = \begin{pmatrix} \mathbf{a}_1^T\mathbf{a}_1 & \mathbf{a}_1^T\mathbf{a}_2 & \cdots & \mathbf{a}_1^T\mathbf{a}_n \\ \mathbf{a}_2^T\mathbf{a}_1 & \mathbf{a}_2^T\mathbf{a}_2 & \cdots & \mathbf{a}_2^T\mathbf{a}_n \\ \vdots & & & \\ \mathbf{a}_n^T\mathbf{a}_1 & \mathbf{a}_n^T\mathbf{a}_2 & \cdots & \mathbf{a}_n^T\mathbf{a}_n \end{pmatrix}$

5. (a) $\begin{pmatrix} 1 & 1 & 1 \\ 2 & 1 & 2 \end{pmatrix} \begin{pmatrix} 4 & -2 & 1 \\ 2 & 3 & 1 \\ 1 & 1 & 2 \end{pmatrix} + \begin{pmatrix} -1 \\ -1 \end{pmatrix} (1 \;\; 2 \;\; 3) = \begin{pmatrix} 6 & 0 & 1 \\ 11 & -1 & 4 \end{pmatrix}$

(c) Let

$$A_{11} = \begin{pmatrix} \frac{3}{5} & -\frac{4}{5} \\ \frac{4}{5} & \frac{3}{5} \end{pmatrix} \qquad A_{12} = \begin{pmatrix} 0 & 0 \\ 0 & 0 \end{pmatrix}$$

$$A_{21} = (0 \quad 0) \qquad\qquad A_{22} = (1 \quad 0)$$

The block multiplication is performed as follows:

$$\begin{pmatrix} A_{11} & A_{12} \\ A_{21} & A_{22} \end{pmatrix} \begin{pmatrix} A_{11}^T & A_{21}^T \\ A_{12}^T & A_{22}^T \end{pmatrix} = \begin{pmatrix} A_{11}A_{11}^T + A_{12}A_{12}^T & A_{11}A_{21}^T + A_{12}A_{22}^T \\ A_{21}A_{11}^T + A_{22}A_{12}^T & A_{21}A_{21}^T + A_{22}A_{22}^T \end{pmatrix}$$

$$= \left(\begin{array}{cc|c} 1 & 0 & 0 \\ 0 & 1 & 0 \\ \hline 0 & 0 & 0 \end{array}\right)$$

6. (a)

$$\begin{aligned} XY^T &= \mathbf{x}_1\mathbf{y}_1^T + \mathbf{x}_2\mathbf{y}_2^T + \mathbf{x}_3\mathbf{y}_3^T \\ &= \begin{pmatrix} 2 \\ 4 \end{pmatrix} \begin{pmatrix} 1 & 2 \end{pmatrix} + \begin{pmatrix} 1 \\ 2 \end{pmatrix} \begin{pmatrix} 2 & 3 \end{pmatrix} + \begin{pmatrix} 5 \\ 3 \end{pmatrix} \begin{pmatrix} 4 & 1 \end{pmatrix} \\ &= \begin{pmatrix} 2 & 4 \\ 4 & 8 \end{pmatrix} + \begin{pmatrix} 2 & 3 \\ 4 & 6 \end{pmatrix} + \begin{pmatrix} 20 & 5 \\ 12 & 3 \end{pmatrix} \end{aligned}$$

(b) Since $\mathbf{y}_i\mathbf{x}_i^T = (\mathbf{x}_i\mathbf{y}_i^T)^T$ for $j = 1, 2, 3$, the outer product expansion of YX^T is just the transpose of the outer product expansion of XY^T. Thus

$$\begin{aligned} YX^T &= \mathbf{y}_1\mathbf{x}_1^T + \mathbf{y}_2\mathbf{x}_2^T + \mathbf{y}_3\mathbf{x}_3^T \\ &= \begin{pmatrix} 2 & 4 \\ 4 & 8 \end{pmatrix} + \begin{pmatrix} 2 & 4 \\ 3 & 6 \end{pmatrix} + \begin{pmatrix} 20 & 12 \\ 5 & 3 \end{pmatrix} \end{aligned}$$

7. It is possible to perform both block multiplications. To see this suppose A_{11} is a $k \times r$ matrix, A_{12} is a $k \times (n-r)$ matrix, A_{21} is an $(m-k) \times r$ matrix and A_{22} is $(m-k) \times (n-r)$. It is possible to perform the block multiplication of AA^T since the matrix multiplication $A_{11}A_{11}^T$, $A_{11}A_{21}^T$, $A_{12}A_{12}^T$, $A_{12}A_{22}^T$, $A_{21}A_{11}^T$, $A_{21}A_{21}^T$, $A_{22}A_{12}^T$, $A_{22}A_{22}^T$ are all possible. It is possible to perform the block multiplication of A^TA since the matrix multiplications $A_{11}^TA_{11}$, $A_{11}^TA_{12}$, $A_{21}^TA_{21}$, $A_{21}^TA_{11}$, $A_{12}^TA_{12}$, $A_{22}^TA_{21}$, $A_{22}^TA_{22}$ are all possible.

8. $AX = A(\mathbf{x}_1, \mathbf{x}_2, \ldots, \mathbf{x}_r) = (A\mathbf{x}_1, A\mathbf{x}_2, \ldots, A\mathbf{x}_r)$
$B = (\mathbf{b}_1, \mathbf{b}_2, \ldots, \mathbf{b}_r)$
$AX = B$ if and only if the column vectors of AX and B are equal

$$A\mathbf{x}_j = \mathbf{b}_j \qquad j = 1, \ldots, r$$

9. (a) Since D is a diagonal matrix, its jth column will have d_{jj} in the jth row and the other entries will all be 0. Thus $\mathbf{d}_j = d_{jj}\mathbf{e}_j$ for $j = 1, \ldots, n$.

(b)

$$\begin{aligned} AD &= A(d_{11}\mathbf{e}_1,\ d_{22}\mathbf{e}_2, \ldots, d_{nn}\mathbf{e}_n) \\ &= (d_{11}A\mathbf{e}_1,\ d_{22}A\mathbf{e}_2, \ldots, d_{nn}A\mathbf{e}_n) \\ &= (d_{11}\mathbf{a}_1,\ d_{22}\mathbf{a}_2, \ldots, d_{nn}\mathbf{a}_n) \end{aligned}$$

10. (a)

$$U\Sigma = \begin{pmatrix} U_1 & U_2 \end{pmatrix} \begin{pmatrix} \Sigma_1 \\ O \end{pmatrix} = U_1\Sigma_1 + U_2O = U_1\Sigma_1$$

(b) If we let $X = U\Sigma$, then

$$X = U_1\Sigma_1 = (\sigma_1\mathbf{u}_1, \sigma_2\mathbf{u}_2, \ldots, \sigma_n\mathbf{u}_n)$$

and it follows that

$$A = U\Sigma V^T = XV^T = \sigma_1\mathbf{u}_1\mathbf{v}_1^T + \sigma_2\mathbf{u}_2\mathbf{v}_2^T + \cdots + \sigma_n\mathbf{u}_n\mathbf{v}_n^T$$

11.

$$\begin{pmatrix} A_{11}^{-1} & C \\ O & A_{22}^{-1} \end{pmatrix}\begin{pmatrix} A_{11} & A_{12} \\ O & A_{22} \end{pmatrix} = \begin{pmatrix} I & A_{11}^{-1}A_{12} + CA_{22} \\ O & I \end{pmatrix}$$

If

$$A_{11}^{-1}A_{12} + CA_{22} = O$$

then

$$C = -A_{11}^{-1}A_{12}A_{22}^{-1}$$

Let

$$B = \begin{pmatrix} A_{11}^{-1} & -A_{11}^{-1}A_{12}A_{22}^{-1} \\ O & A_{22}^{-1} \end{pmatrix}$$

Since $AB = BA = I$ it follows that $B = A^{-1}$.

12. Let $\mathbf{0}$ denote the zero vector in R^n. If A is singular then there exists a vector $\mathbf{x}_1 \neq \mathbf{0}$ such that $A\mathbf{x}_1 = \mathbf{0}$. If we set

$$\mathbf{x} = \begin{pmatrix} \mathbf{x}_1 \\ \mathbf{0} \end{pmatrix}$$

then

$$M\mathbf{x} = \begin{pmatrix} A & C \\ O & B \end{pmatrix}\begin{pmatrix} \mathbf{x}_1 \\ \mathbf{0} \end{pmatrix} = \begin{pmatrix} A\mathbf{x}_1 + C\mathbf{0} \\ O\mathbf{x}_1 + B\mathbf{0} \end{pmatrix} = \begin{pmatrix} \mathbf{0} \\ \mathbf{0} \end{pmatrix}$$

By Theorem 1.4.2, M must be singular. Similarly, if B is singular then there exists a vector $\mathbf{x}_2 \neq \mathbf{0}$ such that $B\mathbf{x}_2 = \mathbf{0}$. So if we set

$$\mathbf{x} = \begin{pmatrix} \mathbf{0} \\ \mathbf{x}_2 \end{pmatrix}$$

then $\mathbf{x}$ is a nonzero vector and $M\mathbf{x}$ is equal to the zero vector.

15. The block form of S^{-1} is given by

$$S^{-1} = \begin{pmatrix} I & -A \\ O & I \end{pmatrix}$$

It follows that

$$\begin{aligned} S^{-1}MS &= \begin{pmatrix} I & -A \\ O & I \end{pmatrix}\begin{pmatrix} AB & O \\ B & O \end{pmatrix}\begin{pmatrix} I & A \\ O & I \end{pmatrix} \\ &= \begin{pmatrix} I & -A \\ O & I \end{pmatrix}\begin{pmatrix} AB & ABA \\ B & BA \end{pmatrix} \end{aligned}$$

$$= \begin{pmatrix} O & O \\ B & BA \end{pmatrix}$$

16. The block multiplication of the two factors yields

$$\begin{pmatrix} I & O \\ B & I \end{pmatrix} \begin{pmatrix} A_{11} & A_{12} \\ O & C \end{pmatrix} = \begin{pmatrix} A_{11} & A_{12} \\ BA_{11} & BA_{12} + C \end{pmatrix}$$

If we equate this matrix with the block form of A and solve for B and C we get

$$B = A_{21}A_{11}^{-1} \qquad \text{and} \qquad C = A_{22} - A_{21}A_{11}^{-1}A_{12}$$

To check that this works note that

$$\begin{aligned} BA_{11} &= A_{21}A_{11}^{-1}A_{11} = A_{21} \\ BA_{12} + C &= A_{21}A_{11}^{-1}A_{12} + A_{22} - A_{21}A_{11}^{-1}A_{12} = A_{22} \end{aligned}$$

and hence

$$\begin{pmatrix} I & O \\ B & I \end{pmatrix} \begin{pmatrix} A_{11} & A_{12} \\ O & C \end{pmatrix} = \begin{pmatrix} A_{11} & A_{12} \\ A_{21} & A_{22} \end{pmatrix} = A$$

17. In order for the block multiplication to work we must have

$$XB = S \qquad \text{and} \qquad YM = T$$

Since both B and M are nonsingular, we can satisfy these conditions by choosing $X = SB^{-1}$ and $Y = TM^{-1}$.

18. (a)

$$BC = \begin{pmatrix} b_1 \\ b_2 \\ \vdots \\ b_n \end{pmatrix} (c) = \begin{pmatrix} b_1c \\ b_2c \\ \vdots \\ b_nc \end{pmatrix} = c\mathbf{b}$$

(b)

$$\begin{aligned} A\mathbf{x} &= (\mathbf{a}_1, \mathbf{a}_2, \ldots, \mathbf{a}_n) \begin{pmatrix} x_1 \\ x_2 \\ \vdots \\ x_n \end{pmatrix} \\ &= \mathbf{a}_1(x_1) + \mathbf{a}_2(x_2) + \cdots + \mathbf{a}_n(x_n) \end{aligned}$$

(c) It follows from parts (a) and (b) that

$$\begin{aligned} A\mathbf{x} &= \mathbf{a}_1(x_1) + \mathbf{a}_2(x_2) + \cdots + \mathbf{a}_n(x_n) \\ &= x_1\mathbf{a}_1 + x_2\mathbf{a}_2 + \cdots + x_n\mathbf{a}_n \end{aligned}$$

19. If $A\mathbf{x} = \mathbf{0}$ for all $\mathbf{x} \in R^n$, then

$$\mathbf{a}_j = A\mathbf{e}_j = \mathbf{0} \quad \text{for} \quad j = 1, \ldots, n$$

and hence A must be the zero matrix.

20. If

$$B\mathbf{x} = C\mathbf{x} \quad \text{for all} \quad \mathbf{x} \in R^n$$

then

$$(B - C)\mathbf{x} = \mathbf{0} \quad \text{for all} \quad \mathbf{x} \in R^n$$

It follows from Exercise 19 that

$$\begin{aligned} B - C &= O \\ B &= C \end{aligned}$$

21. (a)

$$\begin{pmatrix} A^{-1} & \mathbf{0} \\ -\mathbf{c}^T A^{-1} & 1 \end{pmatrix} \begin{pmatrix} A & \mathbf{a} \\ \mathbf{c}^T & \beta \end{pmatrix} \begin{pmatrix} \mathbf{x} \\ x_{n+1} \end{pmatrix} = \begin{pmatrix} A^{-1} & \mathbf{0} \\ -\mathbf{c}^T A^{-1} & 1 \end{pmatrix} \begin{pmatrix} \mathbf{b} \\ b_{n+1} \end{pmatrix}$$

$$\begin{pmatrix} I & A^{-1}\mathbf{a} \\ \mathbf{0}^T & -\mathbf{c}^T A^{-1}\mathbf{a} + \beta \end{pmatrix} \begin{pmatrix} \mathbf{x} \\ x_{n+1} \end{pmatrix} = \begin{pmatrix} A^{-1}\mathbf{b} \\ -\mathbf{c}^T A^{-1}\mathbf{b} + b_{n+1} \end{pmatrix}$$

(b) If

$$\mathbf{y} = A^{-1}\mathbf{a} \qquad \text{and} \qquad \mathbf{z} = A^{-1}\mathbf{b}$$

then

$$(-\mathbf{c}^T\mathbf{y} + \beta)x_{n+1} = -\mathbf{c}^T\mathbf{z} + b_{n+1}$$

$$x_{n+1} = \frac{-\mathbf{c}^T\mathbf{z} + b_{n+1}}{-\mathbf{c}^T\mathbf{y} + \beta} \qquad (\beta - \mathbf{c}^T\mathbf{y} \neq 0)$$

and

$$\mathbf{x} + x_{n+1}A^{-1}\mathbf{a} = A^{-1}\mathbf{b}$$

$$\mathbf{x} = A^{-1}\mathbf{b} - x_{n+1}A^{-1}\mathbf{a} = \mathbf{z} - x_{n+1}\mathbf{y}$$

MATLAB EXERCISES

1. In parts (a), (b), (c) it should turn out that $A1 = A4$ and $A2 = A3$. In part (d) $A1 = A3$ and $A2 = A4$. Exact equality will not occur in parts (c) and (d) because of roundoff error.

2. The solution $\mathbf{x}$ obtained using the \ operation will be more accurate and yield the smaller residual vector. The computation of $\mathbf{x}$ is also more efficient since the solution is computed using Gaussian elimination with partial pivoting and this involves less arithmetic than computing the inverse matrix and multiplying it times $\mathbf{b}$.

3. (a) Since $A\mathbf{x} = \mathbf{0}$ and $\mathbf{x} \neq \mathbf{0}$, it follows from Theorem 1.4.2 that A is singular.

(b) The columns of B are all multiples of $\mathbf{x}$. Indeed,

$$B = (\mathbf{x}, 2\mathbf{x}, 3\mathbf{x}, 4\mathbf{x}, 5\mathbf{x}, 6\mathbf{x})$$

and hence

$$AB = (A\mathbf{x}, 2A\mathbf{x}, 3A\mathbf{x}, 4A\mathbf{x}, 5A\mathbf{x}, 6A\mathbf{x}) = O$$

(c) If $D = B + C$, then

$$AD = AB + AC = O + AC = AC$$

4. By construction B is upper triangular whose diagonal entries are all equal to 1. Thus B is row equivalent to I and hence B is nonsingular. If one changes B by setting $b_{10,1} = -1/256$ and computes $B\mathbf{x}$, the result is the zero vector. Since $\mathbf{x} \neq \mathbf{0}$, the matrix B must be singular.

5. (a) Since A is nonsingular its reduced row echelon form is I. If $E_1, \ldots, E_k$ are elementary matrices such that $E_k \cdots E_1 A = I$, then these same matrices can be used to transform $(A \;\; \mathbf{b})$ to its reduced row echelon form U. It follows then that

$$U = E_k \cdots E_1 (A \;\; \mathbf{b}) = A^{-1}(A \;\; \mathbf{b}) = (I \;\; A^{-1}\mathbf{b})$$

Thus, the last column of U should be equal to the solution $\mathbf{x}$ of the system $A\mathbf{x} = \mathbf{b}$.

(b) After the third column of A is changed, the new matrix A is now singular. Examining the last row of the reduced row echelon form of the augmented matrix $(A \; \mathbf{b})$, we see that the system is inconsistent.

(c) The system $A\mathbf{x} = \mathbf{c}$ is consistent since $\mathbf{y}$ is a solution. There is a free variable x_3, so the system will have infinitely many solutions.

(f) The vector $\mathbf{v}$ is a solution since

$$A\mathbf{v} = A(\mathbf{w} + 3\mathbf{z}) = A\mathbf{w} + 3A\mathbf{z} = \mathbf{c}$$

For this solution the free variable $x_3 = v_3 = 3$. To determine the general solution just set $\mathbf{x} = \mathbf{w} + t\mathbf{z}$. This will give the solution corresponding to $x_3 = t$ for any real number t.

6. (c) There will be no walks of even length from V_i to V_j whenever $i + j$ is odd.

(d) There will be no walks of length k from V_i to V_j whenever $i + j + k$ is odd.

(e) The conjecture is still valid for the graph containing the additional edges.

(f) If the edge $\{V_6, V_8\}$ is included, then the conjecture is no longer valid. There is now a walk of length 1 V_6 to V_8 and $i + j + k = 6 + 8 + 1$ is odd.

8. The change in part (b) should not have a significant effect on the survival potential for the turtles. The change in part (c) will effect the $(2, 2)$ and $(3, 2)$ of the Leslie matrix. The new values for these entries will be $l_{22} = 0.9540$ and $l_{32} = 0.0101$. With these values the Leslie population model should predict that the survival period will double but the turtles will still eventually die out.

9. (b) $\mathbf{x1} = \mathbf{c} - V\mathbf{x2}$.

10. (b)

$$A^{2k} = \begin{pmatrix} I & kB \\ kB & I \end{pmatrix}$$

This can be proved using mathematical induction. In the case $k = 1$

$$A^2 = \begin{pmatrix} O & I \\ I & B \end{pmatrix} \begin{pmatrix} O & I \\ I & B \end{pmatrix} = \begin{pmatrix} I & B \\ B & I \end{pmatrix}$$

If the result holds for $k = m$

$$A^{2m} = \begin{pmatrix} I & mB \\ mB & I \end{pmatrix}$$

then

$$\begin{aligned} A^{2m+2} &= A^2A^{2m} \\ &= \begin{pmatrix} I & B \\ B & I \end{pmatrix} \begin{pmatrix} I & mB \\ mB & I \end{pmatrix} \\ &= \begin{pmatrix} I & (m+1)B \\ (m+1)B & I \end{pmatrix} \end{aligned}$$

It follows by mathematical induction that the result holds for all positive integers k.

(b)

$$A^{2k+1} = AA^{2k} = \begin{pmatrix} O & I \\ I & B \end{pmatrix} \begin{pmatrix} I & kB \\ kB & I \end{pmatrix} = \begin{pmatrix} kB & I \\ I & (k+1)B \end{pmatrix}$$

11. (a) By construction the entries of A were rounded to the nearest integer. The matrix $B = A^TA$ must also have integer entries and it is symmetric since

$$B^T = (A^TA)^T = A^T(A^T)^T = A^TA = B$$

(b)

$$\begin{aligned} LDL^T &= \begin{pmatrix} I & O \\ E & I \end{pmatrix} \begin{pmatrix} B_{11} & O \\ O & F \end{pmatrix} \begin{pmatrix} I & E^T \\ O & I \end{pmatrix} \\ &= \begin{pmatrix} B_{11} & B_{11}E^T \\ EB_{11} & EB_{11}E^T + F \end{pmatrix} \end{aligned}$$

where

$$E = B_{21}B_{11}^{-1} \quad \text{and} \quad F = B_{22} - B_{21}B_{11}^{-1}B_{12}$$

It follows that

$$\begin{aligned} B_{11}E^T &= B_{11}(B_{11}^{-1})^TB_{21}^T = B_{11}B_{11}^{-1}B_{12} = B_{12} \\ EB_{11} &= B_{21}B_{11}^{-1}B_{11} = B_{21} \end{aligned}$$

$$\begin{aligned} EB_{11}E^T + F &= B_{21}E^T + B_{22} - B_{21}B_{11}^{-1}B_{12} \\ &= B_{21}B_{11}^{-1}B_{12} + B_{22} - B_{21}B_{11}^{-1}B_{12} \\ &= B_{22} \end{aligned}$$

Therefore

$$LDL^T = B$$

CHAPTER TEST A

1. The statement is false in general. If the row echelon form has free variables and the linear system is consistent, then there will be infinitely many solutions. However, it is possible to have an inconsistent system whose coefficient matrix will reduce to an echelon form with free variables. For example, if

$$A = \begin{pmatrix} 1 & 1 \\ 0 & 0 \end{pmatrix} \qquad \mathbf{b} = \begin{pmatrix} 1 \\ 1 \end{pmatrix}$$

then A involves one free variable, but the system $A\mathbf{x} = \mathbf{b}$ is inconsistent.

2. The statement is true since the zero vector will always be a solution.

3. The statement is true. A matrix A is nonsingular if and only if it is row equivalent to the I (the identity matrix). A will be row equivalent to I if and only if its reduced row echelon form is I.

4. The statement is false in general. For example, if $A = I$ and $B = -I$, the matrices A and B are both nonsingular, but $A + B = O$ is singular.

5. The statement is false in general. If A and B are nonsingular, then AB must also be nonsingular, however, $(AB)^{-1}$ is equal to $B^{-1}A^{-1}$ rather than $A^{-1}B^{-1}$. For example, if

$$A = \begin{pmatrix} 1 & 1 \\ 0 & 1 \end{pmatrix} \qquad B = \begin{pmatrix} 1 & 0 \\ 1 & 1 \end{pmatrix}$$

then

$$AB = \begin{pmatrix} 2 & 1 \\ 1 & 1 \end{pmatrix} \quad \text{and} \quad (AB)^{-1} = \begin{pmatrix} 1 & -1 \\ -1 & 2 \end{pmatrix}$$

however,

$$A^{-1}B^{-1} = \begin{pmatrix} 1 & -1 \\ 0 & 1 \end{pmatrix} \begin{pmatrix} 1 & 0 \\ -1 & 1 \end{pmatrix} = \begin{pmatrix} 2 & -1 \\ -1 & 1 \end{pmatrix}$$

Note that

$$B^{-1}A^{-1} = \begin{pmatrix} 1 & 0 \\ -1 & 1 \end{pmatrix} \begin{pmatrix} 1 & -1 \\ 0 & 1 \end{pmatrix} = \begin{pmatrix} 1 & -1 \\ -1 & 2 \end{pmatrix} = (AB)^{-1}$$

6. The statement is false in general.

$$(A - B)^2 = A^2 - BA - AB + B^2 \neq A^2 - 2AB + B^2$$

since in general $BA \neq AB$. For example, if

$$A = \begin{pmatrix} 1 & 1 \\ 1 & 1 \end{pmatrix} \quad \text{and} \quad B = \begin{pmatrix} 0 & 1 \\ 0 & 0 \end{pmatrix}$$

then

$$(A - B)^2 = \begin{pmatrix} 1 & 0 \\ 1 & 1 \end{pmatrix}^2 = \begin{pmatrix} 1 & 0 \\ 2 & 1 \end{pmatrix}$$

however,

$$A^2 - 2AB + B^2 = \begin{pmatrix} 2 & 2 \\ 2 & 2 \end{pmatrix} - \begin{pmatrix} 0 & 2 \\ 0 & 2 \end{pmatrix} + \begin{pmatrix} 0 & 0 \\ 0 & 0 \end{pmatrix} = \begin{pmatrix} 2 & 0 \\ 2 & 0 \end{pmatrix}$$

7. The statement is false in general. If A is nonsingular and $AB = AC$, then we can multiply both sides of the equation by A^{-1} and conclude that $B = C$. However, if A is singular, then it is possible to have $AB = AC$ and $B \neq C$. For example, if

$$A = \begin{pmatrix} 1 & 1 \\ 1 & 1 \end{pmatrix}, \quad B = \begin{pmatrix} 1 & 1 \\ 4 & 4 \end{pmatrix}, \quad C = \begin{pmatrix} 2 & 2 \\ 3 & 3 \end{pmatrix}$$

then

$$\begin{aligned} AB &= \begin{pmatrix} 1 & 1 \\ 1 & 1 \end{pmatrix}\begin{pmatrix} 1 & 1 \\ 4 & 4 \end{pmatrix} = \begin{pmatrix} 5 & 5 \\ 5 & 5 \end{pmatrix} \\ AC &= \begin{pmatrix} 1 & 1 \\ 1 & 1 \end{pmatrix}\begin{pmatrix} 2 & 2 \\ 3 & 3 \end{pmatrix} = \begin{pmatrix} 5 & 5 \\ 5 & 5 \end{pmatrix} \end{aligned}$$

8. The statement is false. An elementary matrix is a matrix that is constructed by performing exactly one elementary row operation on the identity matrix. The product of two elementary matrices will be a matrix formed by performing *two* elementary row operations on the identity matrix. For example,

$$E_1 = \begin{pmatrix} 1 & 0 & 0 \\ 2 & 1 & 0 \\ 0 & 0 & 1 \end{pmatrix} \quad \text{and} \quad E_2 = \begin{pmatrix} 1 & 0 & 0 \\ 0 & 1 & 0 \\ 3 & 0 & 1 \end{pmatrix}$$

are elementary matrices, however,

$$E_1E_2 = \begin{pmatrix} 1 & 0 & 0 \\ 2 & 1 & 0 \\ 3 & 0 & 1 \end{pmatrix}$$

is not an elementary matrix.

9. The statement is true. The row vectors of A are $x_1\mathbf{y}^T, x_2\mathbf{y}^T, \ldots, x_n\mathbf{y}^T$. Note, all of the row vectors are multiples of $\mathbf{y}^T$. Since $\mathbf{x}$ and $\mathbf{y}$ are nonzero vectors, at least one of these row vectors must be nonzero. However, if any nonzero row is picked as a pivot row, then since all of the other rows are multiples of the pivot row, they will all be eliminated in the first step of the reduction process. The resulting row echelon form will have exactly one nonzero row.

10. The statement is true. If $\mathbf{b} = \mathbf{a}_1 + \mathbf{a}_2 + \mathbf{a}_3$, then $\mathbf{x} = (1, 1, 1)^T$ is a solution to $A\mathbf{x} = \mathbf{b}$, since

$$A\mathbf{x} = x_1\mathbf{a}_1 + x_2\mathbf{a}_2 + x_3\mathbf{a}_3 = \mathbf{a}_1 + \mathbf{a}_2 + \mathbf{a}_3 = \mathbf{b}$$

If $\mathbf{a}_2 = \mathbf{a}_3$, then we can also express $\mathbf{b}$ as a linear combination

$$\mathbf{b} = \mathbf{a}_1 + 0\mathbf{a}_2 + 2\mathbf{a}_3$$

Thus $\mathbf{y} = (1, 0, 2)^T$ is also a solution to the system. However, if there is more than one solution, then the echelon form of A must involve a free variable. A consistent system with a free variable must have infinitely many solutions.

CHAPTER TEST B

1.

$$\left(\begin{array}{rrrr|r} 1 & -1 & 3 & 2 & 1 \\ -1 & 1 & -2 & 1 & -2 \\ 2 & -2 & 7 & 7 & 1 \end{array}\right) \rightarrow \left(\begin{array}{rrrr|r} 1 & -1 & 3 & 2 & 1 \\ 0 & 0 & 1 & 3 & -1 \\ 0 & 0 & 1 & 3 & -1 \end{array}\right)$$

$$\rightarrow \left(\begin{array}{rrrr|r} 1 & -1 & 0 & -7 & 4 \\ 0 & 0 & 1 & 3 & -1 \\ 0 & 0 & 0 & 0 & 0 \end{array}\right)$$

The free variables are x_2 and x_4. If we set $x_2 = a$ and $x_4 = b$, then

$$x_1 = 4 + a + 7b \qquad \text{and} \qquad x_3 = -1 - 3b$$

and hence the solution set consists of all vectors of the form

$$\mathbf{x} = \begin{pmatrix} 4 + a + 7b \\ a \\ -1 - 3b \\ b \end{pmatrix}$$

2. **(a)** A linear equation in 3 unknowns corresponds to a plane in 3-space.

(b) Given 2 equations in 3 unknowns, each equation corresponds to a plane. If one equation is a multiple of the other then the equations represent the same plane and any point on the that plane will be a solution to the system. If the two planes are distinct then they are either parallel or they intersect in a line. If they are parallel they do not intersect, so the system will have no solutions. If they intersect in a line then there will be infinitely many solutions.

(c) A homogeneous linear system is always consistent since it has the trivial solution $\mathbf{x} = \mathbf{0}$. It follows from part (b) then that a homogeneous system of 2 equations in 3 unknowns must have infinitely many solutions. Geometrically the 2 equations represent planes that both pass through the origin, so if the planes are distinct they must intersect in a line.

3. **(a)** If the system is consistent and there are two distinct solutions there must be a free variable and hence there must be infinitely many solutions. In fact all vectors of the form $\mathbf{x} = \mathbf{x}_1 + c(\mathbf{x}_1 - \mathbf{x}_2)$ will be solutions since

$$A\mathbf{x} = A\mathbf{x}_1 + c(A\mathbf{x}_1 - A\mathbf{x}_2) = \mathbf{b} + c(\mathbf{b} - \mathbf{b}) = \mathbf{b}$$

(b) If we set $\mathbf{z} = \mathbf{x}_1 - \mathbf{x}_2$ then $\mathbf{z} \neq \mathbf{0}$ and $A\mathbf{z} = \mathbf{0}$. Therefore it follows from Theorem 1.4.2 that A must be singular.

4. **(a)** The system will be consistent if and only if the vector $\mathbf{b} = (3, 1)^T$ can be written as a linear combination of the column vectors of A. Linear combinations of the column vectors of A are vectors of the form

$$c_1 \begin{pmatrix} \alpha \\ 2\alpha \end{pmatrix} + c_2 \begin{pmatrix} \beta \\ 2\beta \end{pmatrix} = (c_1\alpha + c_2\beta) \begin{pmatrix} 1 \\ 2 \end{pmatrix}$$

Since $\mathbf{b}$ is not a multiple of $(1, 2)^T$ the system must be inconsistent.

(b) To obtain a consistent system choose $\mathbf{b}$ to be a multiple of $(1,2)^T$. If this is done the second row of the augmented matrix will zero out in the elimination process and you will end up with one equation in 2 unknowns. The reduced system will have infinitely many solutions.

5. **(a)** To transform A to B you need to interchange the second and third rows of A. The elementary matrix that does this is

$$E = \begin{pmatrix} 1 & 0 & 0 \\ 0 & 0 & 1 \\ 0 & 1 & 0 \end{pmatrix}$$

(b) To transform A to C using a column operation you need to subtract twice the second column of A from the first column. The elementary matrix that does this is

$$F = \begin{pmatrix} 1 & 0 & 0 \\ -2 & 1 & 0 \\ 0 & 0 & 1 \end{pmatrix}$$

6. If $\mathbf{b} = 3\mathbf{a}_1 + \mathbf{a}_2 + 4\mathbf{a}_3$ then $\mathbf{b}$ is a linear combination of the column vectors of A and it follows from the consistency theorem that the system $A\mathbf{x} = \mathbf{b}$ is consistent. In fact $\mathbf{x} = (3,1,4)^T$ is a solution to the system.

7. If $\mathbf{a}_1 - 3\mathbf{a}_2 + 2\mathbf{a}_3 = \mathbf{0}$ then $\mathbf{x} = (1,-3,2)^T$ is a solution to $A\mathbf{x} = \mathbf{0}$. It follows from Theorem 1.4.2 that A must be singular.

8. If

$$A = \begin{pmatrix} 1 & 4 \\ 1 & 4 \end{pmatrix} \quad \text{and} \quad B = \begin{pmatrix} 2 & 3 \\ 2 & 3 \end{pmatrix}$$

then

$$A\mathbf{x} = \begin{pmatrix} 1 & 4 \\ 1 & 4 \end{pmatrix}\begin{pmatrix} 1 \\ 1 \end{pmatrix} = \begin{pmatrix} 5 \\ 5 \end{pmatrix} = \begin{pmatrix} 2 & 3 \\ 2 & 3 \end{pmatrix}\begin{pmatrix} 1 \\ 1 \end{pmatrix} = B\mathbf{x}$$

9. In general the product of two symmetric matrices is not necessarily symmetric. For example if

$$A = \begin{pmatrix} 1 & 2 \\ 2 & 2 \end{pmatrix}, \quad B = \begin{pmatrix} 1 & 1 \\ 1 & 4 \end{pmatrix}$$

then A and B are both symmetric but their product

$$AB = \begin{pmatrix} 1 & 2 \\ 2 & 2 \end{pmatrix}\begin{pmatrix} 1 & 1 \\ 1 & 4 \end{pmatrix} = \begin{pmatrix} 3 & 9 \\ 4 & 10 \end{pmatrix}$$

is not symmetric.

10. If E and F are elementary matrices then they are both nonsingular and their inverses are elementary matrices of the same type. If $C = EF$ then C is a product of nonsingular matrices, so C is nonsingular and $C^{-1} = F^{-1}E^{-1}$.

11.

$$A^{-1} = \begin{pmatrix} I & O & O \\ O & I & O \\ O & -B & I \end{pmatrix}$$

12. **(a)** The column partition of A and the row partition of B must match up, so k must be equal to 5. There is really no restriction on r, it can be any integer in the range $1 \leq r \leq 9$. In fact $r = 10$ will work when B has block structure

$$\begin{pmatrix} B_{11} \\ B_{21} \end{pmatrix}$$

(b) The (2,2) block of the product is given by $A_{21}B_{12} + A_{22}B_{22}$

Chapter 2

SECTION 1

1. (c) $\det(A) = -3$

7. Given that $a_{11} = 0$ and $a_{21} \neq 0$, let us interchange the first two rows of A and also multiply the third row through by $-a_{21}$. We end up with the matrix

$$\begin{pmatrix} a_{21} & a_{22} & a_{23} \\ 0 & a_{12} & a_{13} \\ -a_{21}a_{31} & -a_{21}a_{32} & -a_{21}a_{33} \end{pmatrix}$$

Now if we add a_{31} times the first row to the third, we obtain the matrix

$$\begin{pmatrix} a_{21} & a_{22} & a_{23} \\ 0 & a_{12} & a_{13} \\ 0 & a_{31}a_{22} - a_{21}a_{32} & a_{31}a_{23} - a_{21}a_{33} \end{pmatrix}$$

This matrix will be row equivalent to I if and only if

$$\begin{vmatrix} a_{12} & a_{13} \\ a_{31}a_{22} - a_{21}a_{32} & a_{31}a_{23} - a_{21}a_{33} \end{vmatrix} \neq 0$$

Thus the original matrix A will be row equivalent to I if and only if

$$a_{12}a_{31}a_{23} - a_{12}a_{21}a_{33} - a_{13}a_{31}a_{22} + a_{13}a_{21}a_{32} \neq 0$$

8. Theorem 2.1.3. If A is an $n \times n$ triangular matrix then the determinant of A equals the product of the diagonal elements of A.

Proof: The proof is by induction on n. In the case $n = 1$, $A = (a_{11})$ and $\det(A) = a_{11}$. Assume the result holds for all $k \times k$ triangular matrices and let A be a $(k+1) \times (k+1)$ lower triangular matrix. (It suffices to prove the theorem for lower triangular matrices since $\det(A^T) = \det(A)$.) If $\det(A)$ is expanded by cofactors using the first row of A we get

$$\det(A) = a_{11}\det(M_{11})$$

where M_{11} is the $k \times k$ matrix obtained by deleting the first row and column of A. Since M_{11} is lower triangular we have

$$\det(M_{11}) = a_{22}a_{33}\cdots a_{k+1,k+1}$$

and consequently

$$\det(A) = a_{11}a_{22}\cdots a_{k+1,k+1}$$

9. If the ith row of A consists entirely of 0's then

$$\det(A) = a_{i1}A_{i1} + a_{i2}A_{i2} + \cdots + a_{in}A_{in} = 0$$

If the ith column of A consists entirely of 0's then

$$\det(A) = \det(A^T) = 0$$

10. In the case $n = 1$, if A is a matrix of the form

$$\begin{pmatrix} a & b \\ a & b \end{pmatrix}$$

then $\det(A) = ab - ab = 0$. Suppose that the result holds for $(k+1) \times (k+1)$ matrices and that A is a $(k+2) \times (k+2)$ matrix whose ith and jth rows are identical. Expand $\det(A)$ by factors along the mth row where $m \neq i$ and $m \neq j$.

$$\det(A) = a_{m1}\det(M_{m1}) + a_{m2}\det(M_{m2}) + \cdots + a_{m,k+2}\det(M_{m,k+2}).$$

Each M_{ms}, $1 \le s \le k+2$, is a $(k+1) \times (k+1)$ matrix having two rows that are identical. Thus by the induction hypothesis

$$\det(M_{ms}) = 0 \qquad (1 \le s \le k+2)$$

and consequently $\det(A) = 0$.

11. (a) In general $\det(A+B) \neq \det(A) + \det(B)$. For example if

$$A = \begin{pmatrix} 1 & 0 \\ 0 & 0 \end{pmatrix} \quad \text{and} \quad B = \begin{pmatrix} 0 & 0 \\ 0 & 1 \end{pmatrix}$$

then

$$\det(A) + \det(B) = 0 + 0 = 0$$

and

$$\det(A+B) = \det(I) = 1$$

(b)

$$AB = \begin{pmatrix} a_{11}b_{11} + a_{12}b_{21} & a_{11}b_{12} + a_{12}b_{22} \\ a_{21}b_{11} + a_{22}b_{21} & a_{21}b_{12} + a_{22}b_{22} \end{pmatrix}$$

and hence

$$\begin{aligned}\det(AB) &= (a_{11}b_{11}a_{21}b_{12} + a_{11}b_{11}a_{22}b_{22} + a_{12}b_{21}a_{21}b_{12} + a_{12}b_{21}a_{22}b_{22}) \\ &\quad -(a_{21}b_{11}a_{11}b_{12} + a_{21}b_{11}a_{12}b_{22} + a_{22}b_{21}a_{11}b_{12} + a_{22}b_{21}a_{12}b_{22}) \\ &= a_{11}b_{11}a_{22}b_{22} + a_{12}b_{21}a_{21}b_{12} - a_{21}b_{11}a_{12}b_{22} - a_{22}b_{21}a_{11}b_{12}\end{aligned}$$

On the other hand

$$\begin{aligned}\det(A)\det(B) &= (a_{11}a_{22} - a_{21}a_{12})(b_{11}b_{22} - b_{21}b_{12}) \\ &= a_{11}a_{22}b_{11}b_{22} + a_{21}a_{12}b_{21}b_{12} - a_{21}a_{12}b_{11}b_{22} - a_{11}a_{22}b_{21}b_{12}\end{aligned}$$

Therefore $\det(AB) = \det(A)\det(B)$

(c) In part (b) it was shown that for any pair of 2×2 matrices, the determinant of the product of the matrices is equal to the product of the determinants. Thus if A and B are 2×2 matrices, then

$$\det(AB) = \det(A)\det(B) = \det(B)\det(A) = \det(BA)$$

12. (a) If $d = \det(A + B)$, then

$$\begin{aligned}d &= (a_{11} + b_{11})(a_{22} + b_{22}) - (a_{21} + b_{21})(a_{12} + b_{12}) \\ &= a_{11}a_{22} + a_{11}b_{22} + b_{11}a_{22} + b_{11}b_{22} - a_{21}a_{12} - a_{21}b_{12} - b_{21}a_{12} - b_{21}b_{12} \\ &= (a_{11}a_{22} - a_{21}a_{12}) + (b_{11}b_{22} - b_{21}b_{12}) + (a_{11}b_{22} - b_{21}a_{12}) + (b_{11}a_{22} - a_{21}b_{12}) \\ &= \det(A) + \det(B) + \det(C) + \det(D)\end{aligned}$$

(b) If

$$B = EA = \begin{pmatrix} \alpha a_{21} & \alpha a_{22} \\ \beta a_{11} & \beta a_{12} \end{pmatrix}$$

then

$$C = \begin{pmatrix} a_{11} & a_{12} \\ \beta a_{11} & \beta a_{12} \end{pmatrix} \qquad D = \begin{pmatrix} \alpha a_{21} & \alpha a_{22} \\ a_{21} & a_{22} \end{pmatrix}$$

and hence

$$\det(C) = \det(D) = 0$$

It follows from part (a) that

$$\det(A + B) = \det(A) + \det(B)$$

13. Expanding $\det(A)$ by cofactors using the first row we get

$$\det(A) = a_{11}\det(M_{11}) - a_{12}\det(M_{12})$$

If the first row and column of M_{12} are deleted the resulting matrix will be the matrix B obtained by deleting the first two rows and columns of A. Thus if $\det(M_{12})$ is expanded along the first column we get

$$\det(M_{12}) = a_{21}\det(B)$$

Since $a_{21} = a_{12}$ we have

$$\det(A) = a_{11}\det(M_{11}) - a_{12}^2\det(B)$$

SECTION 2

5. To transform the matrix A into the matrix αA one must perform row operation II n times. Each time row operation II is performed the value of the determinant is changed by a factor of α. Thus

$$\det(\alpha A) = \alpha^n \det(A)$$

Alternatively, one can show this result holds by noting that $\det(\alpha I)$ is equal to the product of its diagonal entries. Thus, $\det(\alpha I) = \alpha^n$ and it follows that

$$\det(\alpha A) = \det(\alpha I A) = \det(\alpha I)\det(A) = \alpha^n \det(A)$$

6. Since

$$\det(A^{-1})\det(A) = \det(A^{-1}A) = \det(I) = 1$$

it follows that

$$\det(A^{-1}) = \frac{1}{\det(A)}$$

8. If E is an elementary matrix of type I or II then E is symmetric, so $E^T = E$. If E is an elementary matrix of type III formed from the identity matrix by adding c times its ith row to its jth row, then E^T will be the elementary matrix of type III formed from the identity matrix by adding c times its jth row to its ith row

9. (b) 18; (d) −6; (f) −3

10. Row operation III has no effect on the value of the determinant. Thus if B can be obtained from A using only row operation III, then $\det(B) = \det(A)$. Row operation I has the effect of changing the sign of the determinant. If B is obtained from A using only row operations I and III, then $\det(B) = \det(A)$ if row operation I has been applied an even number of times and $\det(B) = -\det(A)$ if row operation I has been applied an odd number of times.

11. Since $\det(A^2) = \det(A)^2$ it follows that $\det(A^2)$ must be a nonnegative real number. (We are assuming the entries of A are all real numbers.) If $A^2 + I = O$ then $A^2 = -I$ and hence $\det(A^2) = \det(-I)$. This is not possible if n is odd, since for n odd, $\det(-I) = -1$. On the other hand it is possible for $A^2 + I = O$ when n is even. For example when $n = 2$, if we take

$$A = \begin{pmatrix} 0 & 1 \\ -1 & 0 \end{pmatrix}$$

then it is easily verified that $A^2 + I = O$.

12. (a) Row operation III has no effect on the value of the determinant. Thus

$$\det(V) = \begin{vmatrix} 1 & x_1 & x_1^2 \\ 1 & x_2 & x_2^2 \\ 1 & x_3 & x_3^2 \end{vmatrix} = \begin{vmatrix} 1 & x_1 & x_1^2 \\ 0 & x_2 - x_1 & x_2^2 - x_1^2 \\ 0 & x_3 - x_1 & x_3^2 - x_1^2 \end{vmatrix}$$

and hence

$$\begin{aligned}\det(V) &= (x_2 - x_1)(x_3^2 - x_1^2) - (x_3 - x_1)(x_2^2 - x_1^2)\\ &= (x_2 - x_1)(x_3 - x_1)[(x_3 + x_1) - (x_2 + x_1)]\\ &= (x_2 - x_1)(x_3 - x_1)(x_3 - x_2)\end{aligned}$$

(b) The determinant will be nonzero if and only if no two of the x_i values are equal. Thus V will be nonsingular if and only if the three points x_1, x_2, x_3 are distinct.

14. Since

$$\det(AB) = \det(A)\det(B)$$

it follows that $\det(AB) \neq 0$ if and only if $\det(A)$ and $\det(B)$ are both nonzero. Thus AB is nonsingular if and only if A and B are both nonsingular.

15. If $AB = I$, then $\det(AB) = 1$ and hence by Exercise 14 both A and B are nonsingular. It follows then that

$$B = IB = (A^{-1}A)B = A^{-1}(AB) = A^{-1}I = A^{-1}$$

Thus to show that a square matrix A is nonsingular it suffices to show that there exists a matrix B such that $AB = I$. We need not check whether or not $BA = I$.

16. If A is a $n \times n$ skew symmetric matrix, then

$$\det(A) = \det(A^T) = \det(-A) = (-1)^n \det(A)$$

Thus if n is odd then

$$\begin{aligned}\det(A) &= -\det(A)\\ 2\det(A) &= 0\end{aligned}$$

and hence A must be singular.

17. If A_{nn} is nonzero and one subtracts $c = \det(A)/A_{nn}$ from the (n, n) entry of A, then the resulting matrix, call it B, will be singular. To see this look at the cofactor expansion of the B along its last row.

$$\begin{aligned}\det(B) &= b_{n1}B_{n1} + \cdots + b_{n,n-1}B_{n,n-1} + b_{nn}B_{nn}\\ &= a_{n1}A_{n1} + \cdots + A_{n,n-1}A_{n,n-1} + (a_{nn} - c)A_{nn}\\ &= \det(A) - cA_{nn}\\ &= 0\end{aligned}$$

18.

$$X = \begin{pmatrix} x_1 & x_2 & x_3 \\ x_1 & x_2 & x_3 \\ y_1 & y_2 & y_3 \end{pmatrix} \qquad Y = \begin{pmatrix} x_1 & x_2 & x_3 \\ y_1 & y_2 & y_3 \\ y_1 & y_2 & y_3 \end{pmatrix}$$

Since X and Y both have two rows the same it follows that $\det(X) = 0$ and $\det(Y) = 0$. Expanding $\det(X)$ along the first row, we get

$$\begin{aligned}0 &= x_1X_{11} + x_2X_{12} + x_3X_{13}\\ &= x_1z_1 + x_2z_2 + x_3z_3\end{aligned}$$

$$= \mathbf{x}^T\mathbf{z}$$

Expanding $\det(Y)$ along the third row, we get

$$\begin{aligned} 0 &= y_1Y_{31} + y_2Y_{32} + y_3Y_{33} \\ &= y_1z_1 + y_2z_2 + y_3z_3 \\ &= \mathbf{y}^T\mathbf{z}. \end{aligned}$$

19. Prove: Evaluating an $n \times n$ matrix by cofactors requires $(n! - 1)$ additions and

$$\sum_{k=1}^{n-1} \frac{n!}{k!}$$

multiplications.

Proof: The proof is by induction on n. In the case $n = 1$ no additions and multiplications are necessary. Since $1! - 1 = 0$ and

$$\sum_{k=1}^{0} \frac{n!}{k!} = 0$$

the result holds when $n = 1$. Let us assume the result holds when $n = m$. If A is an $(m+1) \times (m+1)$ matrix then

$$\det(A) = a_{11}\det(M_{11}) - a_{12}\det(M_{12}) \pm \cdots \pm a_{1,m+1}\det(M_{1,m+1})$$

Each M_{1j} is an $m \times m$ matrix. By the induction hypothesis the calculation of $\det(M_{1j})$ requires $(m! - 1)$ additions and

$$\sum_{k=1}^{m-1} \frac{m!}{k!}$$

multiplications. The calculation of all $m+1$ of these determinants requires $(m+1)(m! - 1)$ additions and

$$\sum_{k=1}^{m-1} \frac{(m+1)!}{k!}$$

multiplications. The calculation of $\det(A)$ requires an additional $m+1$ multiplications and an additional m additions. Thus the number of additions necessary to compute $\det(A)$ is

$$(m+1)(m! - 1) + m = (m+1)! - 1$$

and the number of multiplications needed is

$$\sum_{k=1}^{m-1} \frac{(m+1)!}{k!} + (m+1) = \sum_{k=1}^{m-1} \frac{(m+1)!}{k!} + \frac{(m+1)!}{m!} = \sum_{k=1}^{m} \frac{(m+1)!}{k!}$$

20. In the elimination method the matrix is reduced to triangular form and the determinant of the triangular matrix is calculated by multiplying its diagonal elements. At the first step of the reduction process the first row is multiplied by $m_{i1} = -a_{i1}/a_{11}$ and then added to the ith row. This requires 1 division,

$n-1$ multiplications and $n-1$ additions. However, this row operation is carried out for $i = 2, \ldots, n$. Thus the first step of the reduction requires $n-1$ divisions, $(n-1)^2$ multiplications and $(n-1)^2$ additions. At the second step of the reduction this same process is carried out on the $(n-1) \times (n-1)$ matrix obtained by deleting the first row and first column of the matrix obtained from step 1. The second step of the elimination process requires $n-2$ divisions, $(n-2)^2$ multiplications, and $(n-2)^2$ additions. After $n-1$ steps the reduction to triangular form will be complete. It will require:

$$\begin{aligned}(n-1)+(n-2)+\cdots+1 &= \frac{n(n-1)}{2} \text{ divisions}\\ (n-1)^2+(n-2)^2+\cdots+1^2 &= \frac{n(2n-1)(n-1)}{6} \text{ multiplications}\\ (n-1)^2+(n-2)^2+\cdots+1^2 &= \frac{n(2n-1)(n-1)}{6} \text{ additions}\end{aligned}$$

It takes $n-1$ additional multiplications to calculate the determinant of the triangular matrix. Thus the calculation $\det(A)$ by the elimination method requires:

$$\frac{n(n-1)}{2} + \frac{n(2n-1)(n-1)}{6} + (n-1) = \frac{(n-1)(n^2+n+3)}{3}$$

multiplications and divisions and $\frac{n(2n-1)(n-1)}{6}$ additions.

SECTION 3

1. (b) $\det(A) = 10$, $\operatorname{adj} A = \begin{pmatrix} 4 & -1 \\ -1 & 3 \end{pmatrix}$, $A^{-1} = \frac{1}{10}\operatorname{adj} A$

(d) $\det(A) = 1$, $A^{-1} = \operatorname{adj} A = \begin{pmatrix} 1 & -1 & 0 \\ 0 & 1 & -1 \\ 0 & 0 & 1 \end{pmatrix}$

6. $A \operatorname{adj} A = O$

7. The solution of $I\mathbf{x} = \mathbf{b}$ is $\mathbf{x} = \mathbf{b}$. It follows from Cramer's rule that

$$b_j = x_j = \frac{\det(B_j)}{\det(I)} = \det(B_j)$$

8. If $\det(A) = \alpha$ then $\det(A^{-1}) = 1/\alpha$. Since $\operatorname{adj} A = \alpha A^{-1}$ we have

$$\det(\operatorname{adj} A) = \det(\alpha A^{-1}) = \alpha^n \det(A^{-1}) = \alpha^{n-1} = \det(A)^{n-1}$$

10. If A is nonsingular then $\det(A) \neq 0$ and hence

$$\operatorname{adj} A = \det(A) A^{-1}$$

is also nonsingular. It follows that

$$(\operatorname{adj} A)^{-1} = \frac{1}{\det(A)}(A^{-1})^{-1} = \det(A^{-1})A$$

Also

$$\operatorname{adj} A^{-1} = \det(A^{-1})(A^{-1})^{-1} = \det(A^{-1})A$$

11. If $A = O$ then $\operatorname{adj} A$ is also the zero matrix and hence is singular. If A is singular and $A \neq O$ then

$$A \operatorname{adj} A = \det(A)I = 0I = O$$

If $\mathbf{a}^T$ is any nonzero row vector of A then

$$\mathbf{a}^T \operatorname{adj} A = \mathbf{0}^T \qquad \text{or} \qquad (\operatorname{adj} A)^T \mathbf{a} = \mathbf{0}$$

By Theorem 1.4.2, $(\operatorname{adj} A)^T$ is singular. Since

$$\det(\operatorname{adj} A) = \det[(\operatorname{adj} A)^T] = 0$$

it follows that adj A is singular.

12. If $\det(A) = 1$ then

$$\operatorname{adj} A = \det(A)A^{-1} = A^{-1}$$

and hence

$$\operatorname{adj}(\operatorname{adj} A) = \operatorname{adj}(A^{-1})$$

It follows from Exercise 10 that

$$\operatorname{adj}(\operatorname{adj} A) = \det(A^{-1})A = \frac{1}{\det(A)}A = A$$

13. The (j, i) entry of Q^T is q_{ij}. Since

$$Q^{-1} = \frac{1}{\det(Q)} \operatorname{adj} Q$$

its (j, i) entry is $Q_{ij}/\det(Q)$. If $Q^{-1} = Q^T$, then

$$q_{ij} = \frac{Q_{ij}}{\det(Q)}$$

MATLAB EXERCISES

2. The magic squares generated by MATLAB have the property that they are nonsingular when n is odd and singular when n is even.

3. (a) The matrix B is formed by interchanging the first two rows of A.
$\det(B) = -\det(A)$.

(b) The matrix C is formed by multiplying the third row of A by 4.
$\det(C) = 4\det(A)$.

(c) The matrix D is formed from A by adding 4 times the fourth row of A to the fifth row.
$\det(D) = \det(A)$.

5. The matrix U is very ill-conditioned. In fact it is singular with respect to the machine precision used by MATLAB. So in general one could not expect to get even a single digit of accuracy in the computed values of $\det(U^T)$ and $\det(UU^T)$. On the other hand, since U is upper triangular, the computed value of $\det(U)$ is the product of its diagonal entries. This value should be accurate to the machine precision.

6. (a) Since $A\mathbf{x} = \mathbf{0}$ and $\mathbf{x} \neq \mathbf{0}$, the matrix must be singular. However, there may be no indication of this if the computations are done in floating point arithmetic. To compute the determinant MATLAB does Gaussian elimination to reduce the matrix to upper triangular form U and then multiplies the diagonal entries of U. In this case the product $u_{11}u_{22}u_{33}u_{44}u_{55}$ has magnitude on the order of 10^{14}. If the computed value of u_{66} has magnitude of the order 10^{-k} and $k \leq 14$, then MATLAB will round the result to a nonzero integer. (MATLAB knows that if you started with an integer matrix, you should end up with an integer value for the determinant.) In general if the determinant is computed in floating point arithmetic, then you cannot expect it to be a reliable indicator of whether or not a matrix is nonsingular.

(c) Since A is singular, $B = AA^T$ should also be singular. Hence the exact value of $\det(B)$ should be 0.

CHAPTER TEST A

1. The statement is true since

$$\det(AB) = \det(A)\det(B) = \det(B)\det(A) = \det(BA)$$

2. The statement is false in general. For example, if

$$A = \begin{pmatrix} 1 & 0 \\ 0 & 0 \end{pmatrix} \quad \text{and} \quad B = \begin{pmatrix} 0 & 0 \\ 0 & 1 \end{pmatrix}$$

then $\det(A + B) = \det(I) = 1$ while $\det(A) + \det(B) = 0 + 0 = 0$.

3. The statement is false in general. For example, if $A = I$, (the 2×2 identity matrix), then $\det(2A) = 4$ while $2\det(A) = 2$.

4. The statement is true. For any matrix C, $\det(C^T) = \det(C)$, so in particular for $C = AB$ we have

$$\det((AB)^T) = \det(AB) = \det(A)\det(B)$$

5. The statement is false in general. For example if

$$A = \begin{pmatrix} 2 & 3 \\ 0 & 4 \end{pmatrix} \quad \text{and} \quad B = \begin{pmatrix} 1 & 0 \\ 0 & 8 \end{pmatrix}$$

then $\det(A) = \det(B) = 8$, however, $A \neq B$.

6. The statement is true. For a product of two matrices we know that

$$\det(AB) = \det(A)\det(B)$$

Using this it is easy to see that the determinant of a product of k matrices is the product of the determinants of the matrices, i.e,

$$\det(A_1A_2\cdots A_k) = \det(A_1)\det(A_2)\cdots\det(A_k)$$

(This can be proved formally using mathematical induction.) In the special case that $A_1 = A_2 = \cdots = A_k$ we have

$$\det(A^k) = \det(A)^k$$

7. The statement is true. A triangular matrix T is nonsingular if and only if

$$\det(T) = t_{11}t_{22}\cdots t_{nn} \neq 0$$

Thus T is nonsingular if and only if all of its diagonal entries are nonzero.

8. The statement is true. If $A\mathbf{x} = \mathbf{0}$ and $\mathbf{x} \neq \mathbf{0}$, then it follows from Theorem 1.4.2 that A must be singular. If A is singular then $\det(A) = 0$.

9. The statement is false in general. For example, if

$$A = \begin{pmatrix} 0 & 1 \\ 1 & 0 \end{pmatrix}$$

and B is the 2×2 identity matrix, then A and B are row equivalent, however, their determinants are not equal.

10. The statement is true. If $A^k = O$, then

$$\det(A)^k = \det(A^k) = \det(O) = 0$$

So $\det(A) = 0$, and hence A must be singular.

CHAPTER TEST B

1. **(a)** $\det(\frac{1}{2}A) = (\frac{1}{2})^3\det(A) = \frac{1}{8}\cdot 4 = \frac{1}{2}$

(b) $\det(B^{-1}A^T) = \det(B^{-1})\det(A^T) = \frac{1}{\det(B)}\det(A) = \frac{1}{6}\cdot 4 = \frac{2}{3}$

(c) $\det(EA^2) = -\det(A^2) = -\det(A)^2 = -16$

2. **(a)**

$$\begin{aligned}\det(A) &= x\begin{vmatrix} x & -1 \\ -1 & x \end{vmatrix} - \begin{vmatrix} 1 & -1 \\ -1 & x \end{vmatrix} + \begin{vmatrix} 1 & x \\ -1 & -1 \end{vmatrix} \\ &= x(x^2-1) - (x-1) + (-1+x) \\ &= x(x-1)(x+1)\end{aligned}$$

(b) The matrix will be singular if x equals 0, 1, or -1.

3. **(a)**

$$\begin{pmatrix} 1 & 1 & 1 & 1 \\ 1 & 2 & 3 & 4 \\ 1 & 3 & 6 & 10 \\ 1 & 4 & 10 & 20 \end{pmatrix} \to \begin{pmatrix} 1 & 1 & 1 & 1 \\ 0 & 1 & 2 & 3 \\ 0 & 2 & 5 & 9 \\ 0 & 3 & 9 & 19 \end{pmatrix} \quad (l_{21} = l_{31} = l_{41} = 1)$$

$$\begin{pmatrix} 1 & 1 & 1 & 1 \\ 0 & 1 & 2 & 3 \\ 0 & 2 & 5 & 9 \\ 0 & 3 & 9 & 19 \end{pmatrix} \rightarrow \begin{pmatrix} 1 & 1 & 1 & 1 \\ 0 & 1 & 2 & 3 \\ 0 & 0 & 1 & 3 \\ 0 & 0 & 3 & 10 \end{pmatrix} \qquad (l_{32} = 2,\ l_{42} = 3)$$

$$\begin{pmatrix} 1 & 1 & 1 & 1 \\ 0 & 1 & 2 & 3 \\ 0 & 0 & 1 & 3 \\ 0 & 0 & 3 & 10 \end{pmatrix} \rightarrow \begin{pmatrix} 1 & 1 & 1 & 1 \\ 0 & 1 & 2 & 3 \\ 0 & 0 & 1 & 3 \\ 0 & 0 & 0 & 1 \end{pmatrix} \qquad (l_{43} = 3)$$

$$A = LU = \begin{pmatrix} 1 & 0 & 0 & 0 \\ 1 & 1 & 0 & 0 \\ 1 & 2 & 1 & 0 \\ 1 & 3 & 3 & 1 \end{pmatrix} \begin{pmatrix} 1 & 1 & 1 & 1 \\ 0 & 1 & 2 & 3 \\ 0 & 0 & 1 & 3 \\ 0 & 0 & 0 & 1 \end{pmatrix}$$

(b) $\det(A) = \det(LU) = \det(L)\det(U) = 1 \cdot 1 = 1$

4. If A is nonsingular then $\det(A) \neq 0$ and it follows that

$$\det(A^TA) = \det(A^T)\det(A) = \det(A)\det(A) = \det(A)^2 > 0$$

Therefore A^TA must be nonsingular.

5. If $B = S^{-1}AS$, then

$$\begin{aligned} \det(B) = \det(S^{-1}AS) &= \det(S^{-1})\det(A)\det(S) \\ &= \frac{1}{\det(S)}\det(A)\det(S) = \det(A) \end{aligned}$$

6. If A is singular then $\det(A) = 0$ and if B is singular then $\det(B)$ so if one of the matrices is singular then

$$\det(C) = \det(AB) = \det(A)\det(B) = 0$$

Therefore the matrix C must be singular.

7. The determinant of $A - \lambda I$ will equal 0 if and only if $A - \lambda I$ is singular. By Theorem 1.4.2, $A - \lambda I$ is singular if and only if there exists a nonzero vector $\mathbf{x}$ such that $(A - \lambda I)\mathbf{x} = \mathbf{0}$. It follows then that $\det(A - \lambda I) = 0$ if and only if $A\mathbf{x} = \lambda\mathbf{x}$ for some $\mathbf{x} \neq \mathbf{0}$.

8. If $A = \mathbf{x}\mathbf{y}^T$ then all of the rows of A are multiples of $\mathbf{y}^T$. In fact $\mathbf{a}(i, :) = x_i\mathbf{y}^T$ for $j = 1, \ldots, n$. It follows that if U is any row echelon form of A then U can have at most one nonzero row. Since A is row equivalent to U and $\det(U) = 0$, it follows that $\det(A) = 0$.

9. Let $\mathbf{z} = \mathbf{x} - \mathbf{y}$. Since $\mathbf{x}$ and $\mathbf{y}$ are distinct it follows that $\mathbf{z} \neq \mathbf{0}$. Since

$$A\mathbf{z} = A\mathbf{x} - A\mathbf{y} = \mathbf{0}$$

it follows from Theorem 1.4.2 that A must be singular and hence $\det(A) = 0$.

10. If A has integer entries then adj A will have integer entries. So if $|\det(A)| = 1$ then

$$A^{-1} = \frac{1}{\det(A)}\operatorname{adj} A = \pm \operatorname{adj} A$$

and hence A^{-1} must also have integer entries.

CHAPTER 3

SECTION 1

3. To show that C is a vector space we must show that all eight axioms are satisfied.

A1. $(a+bi)+(c+di) = (a+c)+(b+d)i$
$$= (c+a)+(d+b)i$$
$$= (c+di)+(a+bi)$$

A2. $(\mathbf{x}+\mathbf{y})+\mathbf{z} = [(x_1+x_2i)+(y_1+y_2i)]+(z_1+z_2i)$
$$= (x_1+y_1+z_1)+(x_2+y_2+z_2)i$$
$$= (x_1+x_2i)+[(y_1+y_2i)+(z_1+z_2i)]$$
$$= \mathbf{x}+(\mathbf{y}+\mathbf{z})$$

A3. $(a+bi)+(0+0i) = (a+bi)$

A4. If $\mathbf{z} = a+bi$ then define $-\mathbf{z} = -a-bi$. It follows that

$$\mathbf{z}+(-\mathbf{z}) = (a+bi)+(-a-bi) = 0+0i = \mathbf{0}$$

A5. $\alpha[(a+bi)+(c+di)] = (\alpha a+\alpha c)+(\alpha b+\alpha d)i$
$$= \alpha(a+bi)+\alpha(c+di)$$

A6. $(\alpha+\beta)(a+bi) = (\alpha+\beta)a+(\alpha+\beta)bi$
$$= \alpha(a+bi)+\beta(a+bi)$$

A7. $(\alpha\beta)(a+bi) = (\alpha\beta)a+(\alpha\beta)bi$
$$= \alpha(\beta a+\beta bi)$$

A8. $1 \cdot (a+bi) = 1 \cdot a + 1 \cdot bi = a + bi$

4. Let $A = (a_{ij})$, $B = (b_{ij})$ and $C = (c_{ij})$ be arbitrary elements of $R^{m\times n}$.

A1. Since $a_{ij}+b_{ij} = b_{ij}+a_{ij}$ for each i and j it follows that $A+B = B+A$.

A2. Since

$$(a_{ij} + b_{ij}) + c_{ij} = a_{ij} + (b_{ij} + c_{ij})$$

for each i and j it follows that

$$(A+B)+C = A+(B+C)$$

A3. Let O be the $m \times n$ matrix whose entries are all 0. If $M = A+O$ then

$$m_{ij} = a_{ij} + 0 = a_{ij}$$

Therefore $A+O = A$.

A4. Define $-A$ to be the matrix whose ijth entry is $-a_{ij}$. Since

$$a_{ij} + (-a_{ij}) = 0$$

for each i and j it follows that

$$A + (-A) = O$$

A5. Since

$$\alpha(a_{ij} + b_{ij}) = \alpha a_{ij} + \alpha b_{ij}$$

for each i and j it follows that

$$\alpha(A+B) = \alpha A + \alpha B$$

A6. Since

$$(\alpha+\beta)a_{ij} = \alpha a_{ij} + \beta a_{ij}$$

for each i and j it follows that

$$(\alpha+\beta)A = \alpha A + \beta A$$

A7. Since

$$(\alpha\beta)a_{ij} = \alpha(\beta a_{ij})$$

for each i and j it follows that

$$(\alpha\beta)A = \alpha(\beta A)$$

A8. Since

$$1 \cdot a_{ij} = a_{ij}$$

for each i and j it follows that

$$1A = A$$

5. Let f, g and h be arbitrary elements of $C[a,b]$.

A1. For all x in $[a,b]$

$$(f+g)(x) = f(x) + g(x) = g(x) + f(x) = (g+f)(x).$$

Therefore

$$f+g = g+f$$

A2. For all x in $[a, b]$,

$$\begin{aligned}[(f+g)+h](x) &= (f+g)(x)+h(x)\\ &= f(x)+g(x)+h(x)\\ &= f(x)+(g+h)(x)\\ &= [f+(g+h)](x)\end{aligned}$$

Therefore

$$[(f+g)+h] = [f+(g+h)]$$

A3. If $z(x)$ is identically 0 on $[a, b]$, then for all x in $[a, b]$

$$(f+z)(x) = f(x)+z(x) = f(x)+0 = f(x)$$

Thus

$$f+z = f$$

A4. Define $-f$ by

$$(-f)(x) = -f(x) \quad \text{for all } x \text{ in } [a, b]$$

Since

$$(f+(-f))(x) = f(x)-f(x) = 0$$

for all x in $[a, b]$ it follows that

$$f+(-f) = z$$

A5. For each x in $[a, b]$

$$\begin{aligned}[\alpha(f+g)](x) &= \alpha f(x)+\alpha g(x)\\ &= (\alpha f)(x)+(\alpha g)(x)\end{aligned}$$

Thus

$$\alpha(f+g) = \alpha f+\alpha g$$

A6. For each x in $[a, b]$

$$\begin{aligned}[(\alpha+\beta)f](x) &= (\alpha+\beta)f(x)\\ &= \alpha f(x)+\beta f(x)\\ &= (\alpha f)(x)+(\beta f)(x)\end{aligned}$$

Therefore

$$(\alpha+\beta)f = \alpha f+\beta f$$

A7. For each x in $[a, b]$,

$$[(\alpha\beta)f](x) = \alpha\beta f(x) = \alpha[\beta f(x)] = [\alpha(\beta f)](x)$$

Therefore

$$(\alpha\beta)f = \alpha(\beta f)$$

A8. For each x in $[a, b]$

$$1f(x) = f(x)$$

Therefore

$$1f = f$$

6. The proof is exactly the same as in Exercise 5.

9. (a) If $\mathbf{y} = \beta\mathbf{0}$ then

$$\mathbf{y} + \mathbf{y} = \beta\mathbf{0} + \beta\mathbf{0} = \beta(\mathbf{0} + \mathbf{0}) = \beta\mathbf{0} = \mathbf{y}$$

and it follows that

$$\begin{aligned}(\mathbf{y} + \mathbf{y}) + (-\mathbf{y}) &= \mathbf{y} + (-\mathbf{y})\\ \mathbf{y} + [\mathbf{y} + (-\mathbf{y})] &= \mathbf{0}\\ \mathbf{y} + \mathbf{0} &= \mathbf{0}\\ \mathbf{y} &= \mathbf{0}\end{aligned}$$

(b) If $\alpha\mathbf{x} = \mathbf{0}$ and $\alpha \neq 0$ then it follows from part (a), A7 and A8 that

$$\mathbf{0} = \frac{1}{\alpha}\mathbf{0} = \frac{1}{\alpha}(\alpha\mathbf{x}) = \left(\frac{1}{\alpha}\alpha\right)\mathbf{x} = 1\mathbf{x} = \mathbf{x}$$

10. Axiom 6 fails to hold.

$$\begin{aligned}(\alpha + \beta)\mathbf{x} &= ((\alpha + \beta)x_1,\ (\alpha + \beta)x_2)\\ \alpha\mathbf{x} + \beta\mathbf{x} &= ((\alpha + \beta)x_1, 0)\end{aligned}$$

12. A1. $x \oplus y = x \cdot y = y \cdot x = y \oplus x$

A2. $(x \oplus y) \oplus z = x \cdot y \cdot z = x \oplus (y \oplus z)$

A3. Since $x \oplus 1 = x \cdot 1 = x$ for all x, it follows that 1 is the zero vector.

A4. Let

$$-x = -1 \circ x = x^{-1} = \frac{1}{x}$$

It follows that

$$x \oplus (-x) = x \cdot \frac{1}{x} = 1 \quad \text{(the zero vector).}$$

Therefore $\frac{1}{x}$ is the additive inverse of x for the operation $\oplus$.

A5. $\alpha \circ (x \oplus y) = (x \oplus y)^\alpha = (x \cdot y)^\alpha = x^\alpha \cdot y^\alpha$

$\alpha \circ x \oplus \alpha \circ y = x^\alpha \oplus y^\alpha = x^\alpha \cdot y^\alpha$

A6. $(\alpha + \beta) \circ x = x^{(\alpha+\beta)} = x^\alpha \cdot x^\beta$

$\alpha \circ x \oplus \beta \circ x = x^\alpha \oplus x^\beta = x^\alpha \cdot x^\beta$

A7. $(\alpha\beta) \circ x = x^{\alpha\beta}$

$\alpha \circ (\beta \circ x) = \alpha \circ x^\beta = (x^\beta)^\alpha = x^{\alpha\beta}$

A8. $1 \circ x = x^1 = x$

Since all eight axioms hold, R^+ is a vector space under the operations of $\circ$ and $\oplus$.

13. The system is not a vector space. Axioms A3, A4, A5, A6 all fail to hold.

14. Axioms 6 and 7 fail to hold. To see this consider the following example. If $\alpha = 1.5$, $\beta = 1.8$ and $x = 1$, then

$$(\alpha + \beta) \circ x = [\![3.3]\!] \cdot 1 = 3$$

and

$$\alpha \circ x + \beta \circ x = [\![1.5]\!] \cdot 1 + [\![1.8]\!] \cdot 1 = 1 \cdot 1 + 1 \cdot 1 = 2$$

So Axiom 6 fails. Furthermore,

$$(\alpha\beta) \circ x = [\![2.7]\!] \cdot 1 = 2$$

and

$$\alpha \circ (\beta \circ x) = [\![1.5]\!]([\![1.8]\!] \cdot 1) = 1 \cdot (1 \cdot 1) = 1$$

so Axiom 7 also fails to hold.

15. If $\{a_n\}$, $\{b_n\}$, $\{c_n\}$ are arbitrary elements of S, then for each n

$$a_n + b_n = b_n + a_n$$

and

$$a_n + (b_n + c_n) = (a_n + b_n) + c_n$$

Hence

$$\begin{aligned}\{a_n\} + \{b_n\} &= \{b_n\} + \{a_n\} \\ \{a_n\} + (\{b_n\} + \{c_n\}) &= (\{a_n\} + \{b_n\}) + \{c_n\}\end{aligned}$$

so Axioms 1 and 2 hold.

The zero vector is just the sequence $\{0, 0, \ldots\}$ and the additive inverse of $\{a_n\}$ is the sequence $\{-a_n\}$. The last four axioms all hold since

$$\begin{aligned}\alpha(a_n + b_n) &= \alpha a_n + \alpha b_n \\ (\alpha + \beta)a_n &= \alpha a_n + \beta a_n \\ \alpha\beta a_n &= \alpha(\beta a_n) \\ 1a_n &= a_n\end{aligned}$$

for each n. Thus all eight axioms hold and hence S is a vector space.

16. If

$$\begin{aligned}p(x) = a_1 + a_2x + \cdots + a_nx^{n-1} &\leftrightarrow \mathbf{a} = (a_1, a_2, \ldots, a_n)^T \\ q(x) = b_1 + b_2x + \cdots + b_nx^{n-1} &\leftrightarrow \mathbf{b} = (b_1, b_2, \ldots, b_n)^T\end{aligned}$$

then

$$\begin{aligned}\alpha p(x) &= \alpha a_1 + \alpha a_2 x + \cdots + \alpha a_n x^{n-1} \\ \alpha\mathbf{a} &= (\alpha a_1, \alpha a_2, \ldots, \alpha a_n)^T\end{aligned}$$

and

$$\begin{aligned}(p+q)(x) &= (a_1 + b_1) + (a_2 + b_2)x + \cdots + (a_n + b_n)x^{n-1} \\ \mathbf{a} + \mathbf{b} &= (a_1 + b_1, a_2 + b_2, \ldots a_n + b_n)^T\end{aligned}$$

Thus

$$\alpha p \leftrightarrow \alpha\mathbf{a} \qquad \text{and} \qquad p + q \leftrightarrow \mathbf{a} + \mathbf{b}$$

SECTION 2

7. $C^n[a,b]$ is a nonempty subset of $C[a,b]$. If $f \in C^n[a,b]$, then $f^{(n)}$ is continuous. Any scalar multiple of a continuous function is continuous. Thus for any scalar α, the function

$$(\alpha f)^{(n)} = \alpha f^{(n)}$$

is also continuous and hence $\alpha f \in C^n[a,b]$. If f and g are vectors in $C^n[a,b]$ then both have continuous nth derivatives and their sum will also have a continuous nth derivative. Thus $f+g \in C^n[a,b]$ and therefore $C^n[a,b]$ is a subspace of $C[a,b]$.

8. (a) If $B \in S_1$, then $AB = BA$. It follows that

$$A(\alpha B) = \alpha AB = \alpha BA = (\alpha B)A$$

and hence $\alpha B \in S_1$.

If B and C are in S_1, then

$$AB = BA \qquad \text{and} \qquad AC = CA$$

thus

$$A(B+C) = AB + AC = BA + CA = (B+C)A$$

and hence $B + C \in S_1$. Therefore S_1 is a subspace of $R^{2\times 2}$.

(b) If $B \in S_2$, then $AB \neq BA$. However, for the scalar 0, we have

$$0B = O \notin S_2$$

Therefore S_2 is not a subspace. (Also, S_2 is not closed under addition.)

(c) If $B \in S_3$, then $BA = O$. It follows that

$$(\alpha B)A = \alpha(BA) = \alpha O = O$$

Therefore, $\alpha B \in S_3$. If B and C are in S_3, then

$$BA = O \qquad \text{and} \qquad CA = O$$

It follows that

$$(B+C)A = BA + CA = O + O = O$$

Therefore $B + C \in S_3$ and hence S_3 is a subspace of $R^{2\times 2}$.

11 (a) $\mathbf{x} \in \text{Span}(\mathbf{x}_1, \mathbf{x}_2)$ if and only if there exist scalars c_1 and c_2 such that

$$c_1\mathbf{x}_1 + c_2\mathbf{x}_2 = \mathbf{x}$$

Thus $\mathbf{x} \in \text{Span}(\mathbf{x}_1, \mathbf{x}_2)$ if and only if the system $X\mathbf{c} = \mathbf{x}$ is consistent. To determine whether or not the system is consistent we can compute the row echelon form of the augmented matrix $(X \,|\, \mathbf{x})$.

$$\left(\begin{array}{rr|r} -1 & 3 & 2 \\ 2 & 4 & 6 \\ 3 & 2 & 6 \end{array}\right) \rightarrow \left(\begin{array}{rr|r} 1 & -3 & -2 \\ 0 & 1 & 1 \\ 0 & 0 & 1 \end{array}\right)$$

The system is inconsistent and therefore $\mathbf{x} \notin \text{Span}(\mathbf{x}_1, \mathbf{x}_2)$.

(b)

$$\left(\begin{array}{rr|r} -1 & 3 & -9 \\ 2 & 4 & -2 \\ 3 & 2 & 5 \end{array}\right) \rightarrow \left(\begin{array}{rr|r} 1 & -3 & -2 \\ 0 & 1 & -2 \\ 0 & 0 & 0 \end{array}\right)$$

The system is consistent and therefore $\mathbf{y} \in \text{Span}(\mathbf{x}_1, \mathbf{x}_2)$.

12. (a) Since the vectors $\mathbf{x}_1, \mathbf{x}_2, \ldots, \mathbf{x}_k$ span V, any vector $\mathbf{v}$ in V can be written as a linear combination $\mathbf{v} = c_1\mathbf{x}_1 + c_2\mathbf{x}_2 + \cdots + c_k\mathbf{x}_k$. If we add a vector $\mathbf{x}_{k+1}$ to our spanning set, then we can write $\mathbf{v}$ as a letter combination of the vectors in this augmented set since

$$\mathbf{v} = c_1\mathbf{x}_1 + c_2\mathbf{x}_2 + \cdots + c_k\mathbf{x}_k + 0\mathbf{v}_{k+1}$$

So the new set of $k+1$ vectors will still be a spanning set.

(b) If one of the vectors, say $\mathbf{x}_k$, is deleted from the set then we may or may not end up with a spanning set. It depends on whether $\mathbf{x}_k$ is in $\text{Span}(\mathbf{x}_1, \mathbf{x}_2, \ldots, \mathbf{x}_{k-1})$. If $\mathbf{x}_k \notin \text{Span}(\mathbf{x}_1, \mathbf{x}_2, \ldots, \mathbf{x}_{k-1})$, then $\{\mathbf{x}_1, \mathbf{x}_2, \ldots, \mathbf{x}_{k-1}\}$ cannot be a spanning set. On the other hand if $\mathbf{x}_k \in \text{Span}(\mathbf{x}_1, \mathbf{x}_2, \ldots, \mathbf{x}_{k-1})$, then

$$\text{Span}(\mathbf{x}_1, \mathbf{x}_2, \ldots, \mathbf{x}_k) = \text{Span}(\mathbf{x}_1, \mathbf{x}_2, \ldots, \mathbf{x}_{k-1})$$

and hence the $k-1$ vectors will span the entire vector space.

13. If $A = (a_{ij})$ is any element of $R^{2\times 2}$, then

$$\begin{aligned} A &= \begin{pmatrix} a_{11} & 0 \\ 0 & 0 \end{pmatrix} + \begin{pmatrix} 0 & a_{12} \\ 0 & 0 \end{pmatrix} + \begin{pmatrix} 0 & 0 \\ a_{21} & 0 \end{pmatrix} + \begin{pmatrix} 0 & 0 \\ 0 & a_{22} \end{pmatrix} \\ &= a_{11}E_{11} + a_{12}E_{12} + a_{21}E_{21} + a_{22}E_{22} \end{aligned}$$

15. If $\{a_n\} \in S_0$, then $a_n \to 0$ as $n \to \infty$. If α is any scalar, then $\alpha a_n \to 0$ as $n \to \infty$ and hence $\{\alpha a_n\} \in S_0$. If $\{b_n\}$ is also an element of S_0, then $b_n \to 0$ as $n \to \infty$ and it follows that

$$\lim_{n\to\infty}(a_n + b_n) = \lim_{n\to\infty} a_n + \lim_{n\to\infty} b_n = 0 + 0 = 0$$

Therefore $\{a_n + b_n\} \in S_0$, and it follows that S_0 is a subspace of S.

16. Let $S \neq \{0\}$ be a subspace of R^1 and let $\mathbf{a}$ be an arbitrary element of R^1. If $\mathbf{s}$ is a nonzero element of S, then we can define a scalar α to be the real number a/s. Since S is a subspace it follows that

$$\alpha\mathbf{s} = \frac{a}{s}\mathbf{s} = \mathbf{a}$$

is an element of S. Therefore $S = R^1$.

17. (a) implies (b).

If $N(A) = \{\mathbf{0}\}$, then $A\mathbf{x} = \mathbf{0}$ has only the trivial solution $\mathbf{x} = \mathbf{0}$. By Theorem 1.4.2, A must be nonsingular.

(b) implies (c).

If A is nonsingular then $A\mathbf{x} = \mathbf{b}$ if and only if $\mathbf{x} = A^{-1}\mathbf{b}$. Thus $A^{-1}\mathbf{b}$ is the unique solution to $A\mathbf{x} = \mathbf{b}$.

(c) implies (a).

If the equation $A\mathbf{x} = \mathbf{b}$ has a unique solution for each $\mathbf{b}$, then in particular for $\mathbf{b} = \mathbf{0}$ the solution $\mathbf{x} = \mathbf{0}$ must be unique. Therefore $N(A) = \{\mathbf{0}\}$.

18. Let α be a scalar and let $\mathbf{x}$ and $\mathbf{y}$ be elements of $U \cap V$. The vectors $\mathbf{x}$ and $\mathbf{y}$ are elements of both U and V. Since U and V are subspaces it follows that

$$\alpha\mathbf{x} \in U \quad \text{and} \quad \mathbf{x}+\mathbf{y} \in U$$

$$\alpha\mathbf{x} \in V \quad \text{and} \quad \mathbf{x}+\mathbf{y} \in V$$

Therefore

$$\alpha\mathbf{x} \in U \cap V \quad \text{and} \quad \mathbf{x}+\mathbf{y} \in U \cap V$$

Thus $U \cap V$ is a subspace of W.

19. $S \cup T$ is not a subspace of R^2.

$$S \cup T = \{(s,t)^T \mid s = 0 \text{ or } t = 0\}$$

The vectors $\mathbf{e}_1$ and $\mathbf{e}_2$ are both in $S \cup T$, however, $\mathbf{e}_1 + \mathbf{e}_2 \notin S \cup T$.

20. If $\mathbf{z} \in U + V$, then $\mathbf{z} = \mathbf{u} + \mathbf{v}$ where $\mathbf{u} \in U$ and $\mathbf{v} \in V$. Since U and V are subspaces it follows that

$$\alpha\mathbf{u} \in U \quad \text{and} \quad \alpha\mathbf{v} \in V$$

for all scalars α. Thus

$$\alpha\mathbf{z} = \alpha\mathbf{u} + \alpha\mathbf{v}$$

is an element of $U + V$. If $\mathbf{z}_1$ and $\mathbf{z}_2$ are elements of $U + V$, then

$$\mathbf{z}_1 = \mathbf{u}_1 + \mathbf{v}_1 \quad \text{and} \quad \mathbf{z}_2 = \mathbf{u}_2 + \mathbf{v}_2$$

where $\mathbf{u}_1, \mathbf{u}_2 \in U$ and $\mathbf{v}_1, \mathbf{v}_2 \in V$. Since U and V are subspaces it follows that

$$\mathbf{u}_1 + \mathbf{u}_2 \in U \quad \text{and} \quad \mathbf{v}_1 + \mathbf{v}_2 \in V$$

Thus

$$\mathbf{z}_1 + \mathbf{z}_2 = (\mathbf{u}_1 + \mathbf{v}_1) + (\mathbf{u}_2 + \mathbf{v}_2) = (\mathbf{u}_1 + \mathbf{u}_2) + (\mathbf{v}_1 + \mathbf{v}_2)$$

is an element of $U + V$. Therefore $U + V$ is a subspace of W.

21. **(a)** The distributive law does not work in general. For a counterexample, consider the vector space R^2. If we set $\mathbf{y} = \mathbf{e}_1 + \mathbf{e}_2$ and let

$$S = \text{Span}(\mathbf{e}_1), \quad T = \text{Span}(\mathbf{e}_2), \quad U = \text{Span}(\mathbf{y})$$

then

$$T + U = R^2, \quad S \cap T = \{\mathbf{0}\}, \quad S \cap U = \{\mathbf{0}\}$$

and hence

$$\begin{aligned} S \cap (T + U) &= S \cap R^2 = S \\ (S \cap T) + (S \cap U) &= \{\mathbf{0}\} + \{\mathbf{0}\} = \{\mathbf{0}\} \end{aligned}$$

(b) This distributive law also does not work in general. For a counterexample we can use the same subspaces S, T, and U of R^2 that were used in part (a). Since

$$T \cap U = \{\mathbf{0}\} \qquad \text{and} \qquad S + U = R^2$$

it follows that

$$\begin{aligned} S + (T \cap U) &= S + \{\mathbf{0}\} = S \\ (S + T) \cap (S + U) &= R^2 \cap R^2 = R^2 \end{aligned}$$

SECTION 3

5. (a) If $\mathbf{x}_{k+1} \in \text{Span}(\mathbf{x}_1, \mathbf{x}_2, \ldots, \mathbf{x}_k)$, then the new set of vectors will be linearly dependent. To see this suppose that

$$\mathbf{x}_{k+1} = c_1\mathbf{x}_1 + c_2\mathbf{x}_2 + \cdots + c_k\mathbf{x}_k$$

If we set $c_{k+1} = -1$, then

$$c_1\mathbf{x}_1 + c_2\mathbf{x}_2 + \cdots + c_k\mathbf{x}_k + c_{k+1}\mathbf{x}_{k+1} = \mathbf{0}$$

with at least one of the coefficients, namely c_{k+1}, being nonzero.
On the other hand if $\mathbf{x}_{k+1} \notin \text{Span}(\mathbf{x}_1, \mathbf{x}_2, \ldots, \mathbf{x}_k)$ and

$$c_1\mathbf{x}_1 + c_2\mathbf{x}_2 + \cdots + c_k\mathbf{x}_k + c_{k+1}\mathbf{x}_{k+1} = \mathbf{0}$$

then $c_{k+1} = 0$ (otherwise we could solve for $\mathbf{x}_{k+1}$ in terms of the other vectors). But then

$$c_1\mathbf{x}_1 + c_2\mathbf{x}_2 + \cdots + c_k\mathbf{x}_k + c_k\mathbf{x}_k = \mathbf{0}$$

and it follows from the independence of $\mathbf{x}_1, \ldots, \mathbf{x}_k$ that all of the c_i coefficients are zero and hence that $\mathbf{x}_1, \ldots, \mathbf{x}_{k+1}$ are linearly independent. Thus if $\mathbf{x}_1, \ldots, \mathbf{x}_k$ are linearly independent and we add a vector $\mathbf{x}_{k+1}$ to the collection, then the new set of vectors will be linearly independent if and only if $\mathbf{x}_{k+1} \notin \text{Span}(\mathbf{x}_1, \mathbf{x}_2, \ldots, \mathbf{x}_k)$

(b) Suppose that $\mathbf{x}_1, \mathbf{x}_2, \ldots, \mathbf{x}_k$ are linearly independent. To test whether or not $\mathbf{x}_1, \mathbf{x}_2, \ldots, \mathbf{x}_{k-1}$ are linearly independent consider the equation

$$(1) \qquad c_1\mathbf{x}_1 + c_2\mathbf{x}_2 + \cdots + c_{k-1}\mathbf{x}_{k-1} = \mathbf{0}$$

If $c_1, c_2, \ldots, c_{k-1}$ work in equation (1), then

$$c_1\mathbf{x}_1 + c_2\mathbf{x}_2 + \cdots + c_{k-1}\mathbf{x}_{k-1} + 0\mathbf{x}_k = \mathbf{0}$$

and it follows from the independence of $\mathbf{x}_1, \ldots, \mathbf{x}_k$ that

$$c_1 = c_2 = \cdots = c_{k-1} = 0$$

and hence $\mathbf{x}_1, \ldots, \mathbf{x}_{k-1}$ must be linearly independent.

7. (a) $W(\cos \pi x, \sin \pi x) = \pi$. Since the Wronskian is not identically zero the vectors are linearly independent.

(b) $W(x, e^x, e^{2x}) = 2(x - 1)e^{3x} \neq 0$

(c) $W(x^2, \ln(1+x^2), 1+x^2) = \dfrac{-8x^3}{(1+x^2)^2} \not\equiv 0$

(d) To see that x^3 and $|x|^3$ are linearly independent suppose

$$c_1 x^3 + c_2 |x|^3 \equiv 0$$

on $[-1, 1]$. Setting $x = 1$ and $x = -1$ we get

$$\begin{aligned} c_1 + c_2 &= 0 \\ -c_1 + c_2 &= 0 \end{aligned}$$

The only solution to this system is $c_1 = c_2 = 0$. Thus x^3 and $|x|^3$ are linearly independent.

8. The vectors are linearly dependent since

$$\cos x - 1 + 2\sin^2 \frac{x}{2} \equiv 0$$

on $[-\pi, \pi]$.

10. (a) If

$$c_1(2x) + c_2|x| = 0$$

for all x in $[-1, 1]$, then in particular we have

$$\begin{aligned} -2c_1 + c_2 &= 0 \qquad (x = -1) \\ 2c_1 + c_2 &= 0 \qquad (x = 1) \end{aligned}$$

and hence $c_1 = c_2 = 0$. Therefore $2x$ and $|x|$ are linearly independent in $C[-1, 1]$.

(b) For all x in $[0, 1]$

$$1 \cdot 2x + (-2)|x| = 0$$

Therefore $2x$ and $|x|$ are linearly dependent in $C[0, 1]$.

11. Let $\mathbf{v}_1, \ldots, \mathbf{v}_n$ be vectors in a vector space V. If one of the vectors, say $\mathbf{v}_1$, is the zero vector then set

$$c_1 = 1, \quad c_2 = c_3 = \cdots = c_n = 0$$

Since

$$c_1\mathbf{v}_1 + c_2\mathbf{v}_2 + \cdots + c_n\mathbf{v}_n = \mathbf{0}$$

and $c_1 \neq 0$, it follows that $\mathbf{v}_1, \ldots, \mathbf{v}_n$ are linearly dependent.

12. If $\mathbf{v}_1 = \alpha\mathbf{v}_2$, then

$$1\mathbf{v}_1 - \alpha\mathbf{v}_2 = \mathbf{0}$$

and hence $\mathbf{v}_1, \mathbf{v}_2$ are linearly dependent. Conversely, if $\mathbf{v}_1$, $\mathbf{v}_2$ are linearly dependent, then there exists scalars c_1, c_2, not both zero, such that

$$c_1\mathbf{v}_1 + c_2\mathbf{v}_2 = \mathbf{0}$$

If say $c_1 \neq 0$, then

$$\mathbf{v}_1 = -\frac{c_2}{c_1}\mathbf{v}_2$$

13. Let $\mathbf{v}_1, \mathbf{v}_2, \ldots, \mathbf{v}_n$ be a linearly independent set of vectors and suppose there is a subset, say $\mathbf{v}_1, \ldots, \mathbf{v}_k$ of linearly dependent vectors. This would imply that there exist scalars $c_1, c_2, \ldots, c_k$, not all zero, such that

$$c_1\mathbf{v}_1 + c_2\mathbf{v}_2 + \cdots + c_k\mathbf{v}_k = \mathbf{0}$$

but then

$$c_1\mathbf{v}_1 + \cdots + c_k\mathbf{v}_k + 0\mathbf{v}_{k+1} + \cdots + 0\mathbf{v}_n = \mathbf{0}$$

This contradicts the original assumption that $\mathbf{v}_1, \mathbf{v}_2, \ldots, \mathbf{v}_n$ are linearly independent.

14. If $\mathbf{x} \in N(A)$ then $A\mathbf{x} = \mathbf{0}$. Partitioning A into columns and $\mathbf{x}$ into rows and performing the block multiplication, we get

$$x_1\mathbf{a}_1 + x_2\mathbf{a}_2, \cdots + x_n\mathbf{a}_n = \mathbf{0}$$

Since $\mathbf{a}_1, \mathbf{a}_2, \ldots, \mathbf{a}_n$ are linearly independent it follows that

$$x_1 = x_2 = \cdots = x_n = 0$$

Therefore $\mathbf{x} = \mathbf{0}$ and hence $N(A) = \{\mathbf{0}\}$.

15. If

$$c_1\mathbf{y}_1 + c_2\mathbf{y}_2 + \cdots + c_k\mathbf{y}_k = \mathbf{0}$$

then

$$\begin{aligned} c_1A\mathbf{x}_1 + c_2A\mathbf{x}_2 + \cdots + c_kA\mathbf{x}_k &= \mathbf{0} \\ A(c_1\mathbf{x}_1 + c_2\mathbf{x}_2 + \cdots + c_k\mathbf{x}_k) &= \mathbf{0} \end{aligned}$$

Since A is nonsingular it follows that

$$c_1\mathbf{x}_1 + c_2\mathbf{x}_2 + \cdots + c_k\mathbf{x}_k = \mathbf{0}$$

and since $\mathbf{x}_1, \ldots, \mathbf{x}_k$ are linearly independent it follows that

$$c_1 = c_2 = \cdots = c_k = 0$$

Therefore $\mathbf{y}_1, \mathbf{y}_2, \ldots, \mathbf{y}_k$ are linearly independent.

16. Since $\mathbf{v}_1, \ldots, \mathbf{v}_n$ span V we can write

$$\mathbf{v} = c_1\mathbf{v}_1 + c_2\mathbf{v}_2 + \cdots + c_n\mathbf{v}_n$$

If we set $c_{n+1} = -1$ then $c_{n+1} \neq 0$ and

$$c_1\mathbf{v}_1 + \cdots + c_n\mathbf{v}_n + c_{n+1}\mathbf{v} = \mathbf{0}$$

Thus $\mathbf{v}_1, \ldots, \mathbf{v}_n$, $\mathbf{v}$ are linearly dependent.

17. If $\{\mathbf{v}_2, \ldots, \mathbf{v}_n\}$ were a spanning set for V then we could write

$$\mathbf{v}_1 = c_2\mathbf{v}_2 + \cdots + c_n\mathbf{v}_n$$

Setting $c_1 = -1$, we would have

$$c_1\mathbf{v}_1 + c_2\mathbf{v}_2 + \cdots + c_n\mathbf{v}_n = \mathbf{0}$$

which would contradict the linear independence of $\mathbf{v}_1, \mathbf{v}_2, \ldots, \mathbf{v}_n$.

SECTION 4

3. (a) Since

$$\begin{vmatrix} 2 & 4 \\ 1 & 3 \end{vmatrix} = 2 \neq 0$$

it follows that $\mathbf{x}_1$ and $\mathbf{x}_2$ are linearly independent and hence form a basis for R^2.

(b) It follows from Theorem 3.4.1 that any set of more than two vectors in R^2 must be linearly dependent.

5. (a) Since

$$\begin{vmatrix} 2 & 3 & 2 \\ 1 & -1 & 6 \\ 3 & 4 & 4 \end{vmatrix} = 0$$

it follows that $\mathbf{x}_1$, $\mathbf{x}_2$, $\mathbf{x}_3$ are linearly dependent.

(b) If $c_1\mathbf{x}_1 + c_2\mathbf{x}_2 = \mathbf{0}$, then

$$\begin{aligned} 2c_1 + 3c_2 &= 0 \\ c_1 - c_2 &= 0 \\ 3c_1 + 4c_2 &= 0 \end{aligned}$$

and the only solution to this system is $c_1 = c_2 = 0$. Therefore $\mathbf{x}_1$ and $\mathbf{x}_2$ are linearly independent.

8 (a) Since the dimension of R^3 is 3, it takes at least three vectors to span R^3. Therefore $\mathbf{x}_1$ and $\mathbf{x}_2$ cannot possibly span R^3.

(b) The matrix X must be nonsingular or satisfy an equivalent condition such as $\det(X) \neq 0$.

(c) If $\mathbf{x}_3 = (a, b, c)^T$ and $X = (\mathbf{x}_1, \mathbf{x}_2, \mathbf{x}_3)$ then

$$\det(X) = \begin{vmatrix} 1 & 3 & a \\ 1 & -1 & b \\ 1 & 4 & c \end{vmatrix} = 5a - b - 4c$$

If one chooses a, b, and c so that

$$5a - b - 4c \neq 0$$

then $\{\mathbf{x}_1, \mathbf{x}_2, \mathbf{x}_3\}$ will be a basis for R^3.

9. (a) If $\mathbf{a}_1$ and $\mathbf{a}_2$ are linearly independent then they span a 2-dimensional subspace of R^3. A 2-dimensional subspace of R^3 corresponds to a plane through the origin in 3-space.

(b) If $\mathbf{b} = A\mathbf{x}$ then

$$\mathbf{b} = x_1\mathbf{a}_1 + x_2\mathbf{a}_2$$

so $\mathbf{b}$ is in $\text{Span}(\mathbf{a}_1, \mathbf{a}_2)$ and hence the dimension of $\text{Span}(\mathbf{a}_1, \mathbf{a}_2, \mathbf{b})$ is 2.

10. We must find a subset of three vectors that are linearly independent. Clearly $\mathbf{x}_1$ and $\mathbf{x}_2$ are linearly independent, but

$$\mathbf{x}_3 = \mathbf{x}_2 - \mathbf{x}_1$$

so $\mathbf{x}_1$, $\mathbf{x}_2$, $\mathbf{x}_3$ are linearly dependent. Consider next the vectors $\mathbf{x}_1$, $\mathbf{x}_2$, $\mathbf{x}_4$. If $X = (\mathbf{x}_1, \mathbf{x}_2, \mathbf{x}_4)$ then

$$\det(X) = \begin{vmatrix} 1 & 2 & 2 \\ 2 & 5 & 7 \\ 2 & 4 & 4 \end{vmatrix} = 0$$

so these three vectors are also linearly dependent. Finally if use $\mathbf{x}_5$ and form the matrix $X = (\mathbf{x}_1, \mathbf{x}_2, \mathbf{x}_5)$ then

$$\det(X) = \begin{vmatrix} 1 & 2 & 1 \\ 2 & 5 & 1 \\ 2 & 4 & 0 \end{vmatrix} = -2$$

so the vectors $\mathbf{x}_1$, $\mathbf{x}_2$, $\mathbf{x}_5$ are linearly independent and hence form a basis for R^3.

16. $\dim U = 2$. The set $\{\mathbf{e}_1, \mathbf{e}_2\}$ is a basis for U.
$\dim V = 2$. The set $\{\mathbf{e}_2, \mathbf{e}_3\}$ is a basis for V.
$\dim U \cap V = 1$. The set $\{\mathbf{e}_2\}$ is a basis for $U \cap V$.
$\dim U + V = 3$. The set $\{\mathbf{e}_1, \mathbf{e}_2, \mathbf{e}_3\}$ is a basis for $U + V$.

17. Let $\{\mathbf{u}_1, \mathbf{u}_2\}$ be a basis for U and $\{\mathbf{v}_1, \mathbf{v}_2\}$ be a basis for V. It follows from Theorem 3.4.1 that $\mathbf{u}_1, \mathbf{u}_2, \mathbf{v}_1, \mathbf{v}_2$ are linearly dependent. Thus there exist scalars c_1, c_2, c_3, c_4 not all zero such that

$$c_1\mathbf{u}_1 + c_2\mathbf{u}_2 + c_3\mathbf{v}_1 + c_4\mathbf{v}_2 = \mathbf{0}$$

Let

$$\mathbf{x} = c_1\mathbf{u}_1 + c_2\mathbf{u}_2 = -c_3\mathbf{v}_1 - c_4\mathbf{v}_2$$

The vector $\mathbf{x}$ is an element of $U \cap V$. We claim $\mathbf{x} \neq \mathbf{0}$, for if $\mathbf{x} = \mathbf{0}$, then

$$c_1\mathbf{u}_1 + c_2\mathbf{u}_2 = \mathbf{0} = -c_3\mathbf{v}_1 - c_4\mathbf{v}_2$$

and by the linear independence of $\mathbf{u}_1$ and $\mathbf{u}_2$ and the linear independence of $\mathbf{v}_1$ and $\mathbf{v}_2$ we would have

$$c_1 = c_2 = c_3 = c_4 = 0$$

contradicting the definition of the c_i's.

18. Let U and V be subspaces of R^n with the property that $U \cap V = \{\mathbf{0}\}$. If either $U = \{\mathbf{0}\}$ or $V = \{\mathbf{0}\}$ the result is obvious, so assume that both subspaces are nontrivial with $\dim U = k > 0$ and $\dim V = r > 0$. Let $\{\mathbf{u}_1, \ldots, \mathbf{u}_k\}$ be a basis for U and let $\{\mathbf{v}_1, \ldots, \mathbf{v}_r\}$ be a basis for V. The vectors $\mathbf{u}_1, \ldots, \mathbf{u}_k, \mathbf{v}_1, \ldots, \mathbf{v}_r$ span $U + V$. We claim that these vectors form a basis for $U + V$ and hence that $\dim U + \dim V = k + r$. To show this we must show that the vectors are linearly independent. Thus we must show that if

$$c_1\mathbf{u}_1 + \cdots + c_k\mathbf{u}_k + c_{k+1}\mathbf{v}_1 + \cdots + c_{k+r}\mathbf{v}_r = \mathbf{0} \tag{2}$$

then $c_1 = c_2 = \cdots = c_{k+r} = 0$. If we set

$$\mathbf{u} = c_1\mathbf{u}_1 + \cdots + c_k\mathbf{u}_k \qquad \text{and} \qquad \mathbf{v} = c_{k+1}\mathbf{v}_1 + \cdots + c_{k+r}\mathbf{v}_r$$

then equation (2) becomes

$$\mathbf{u} + \mathbf{v} = \mathbf{0}$$

This implies $\mathbf{u} = -\mathbf{v}$ and hence that both $\mathbf{u}$ and $\mathbf{v}$ are in both in $U \cap V = \{\mathbf{0}\}$. Thus we have

$$\begin{aligned} \mathbf{u} &= c_1\mathbf{u}_1 + \cdots + c_k\mathbf{u}_k = \mathbf{0} \\ \mathbf{v} &= c_{k+1}\mathbf{v}_1 + \cdots + c_{k+r}\mathbf{v}_r = \mathbf{0} \end{aligned}$$

So, by the independence of $\mathbf{u}_1, \ldots, \mathbf{u}_k$ and the independence of $\mathbf{v}_1, \ldots, \mathbf{v}_r$ it follows that

$$c_1 = c_2 = \cdots = c_{k+r} = 0$$

SECTION 5

11. The transition matrix from E to F is $U^{-1}V$. To compute $U^{-1}V$, note that

$$U^{-1}(U \mid V) = (I \mid U^{-1}V)$$

and hence $(I \mid U^{-1}V)$ and $(U \mid V)$ are row equivalent. Thus $(I \mid U^{-1}V)$ is the reduced row echelon form of $(U \mid V)$.

SECTION 6

1. (a) The reduced row echelon form of the matrix is

$$\begin{pmatrix} 1 & 0 & 2 \\ 0 & 1 & 0 \\ 0 & 0 & 0 \end{pmatrix}$$

Thus $(1,\ 0,\ 2)$ and $(0,\ 1,\ 0)$ form a basis for the row space. The first and second columns of the original matrix form a basis for the column space:

$$\mathbf{a}_1 = (1,\ 2,\ 4)^T \quad \text{and} \quad \mathbf{a}_2 = (3,\ 1,\ 7)^T$$

The reduced row echelon form involves one free variable and hence the nullspace will have dimension 1. Setting $x_3 = 1$, we get $x_1 = -2$ and $x_2 = 0$. Thus $(-2,\ 0,\ 1)^T$ is a basis for the nullspace.

(b) The reduced row echelon form of the matrix is

$$\begin{pmatrix} 1 & 0 & 0 & -10/7 \\ 0 & 1 & 0 & -2/7 \\ 0 & 0 & 1 & 0 \end{pmatrix}$$

Clearly then, the set

$$\{(1,\ 0,\ 0,\ -10/7),\ (0,\ 1,\ 0,\ -2/7),\ (0,\ 0,\ 1,\ 0)\}$$

is a basis for the row space. Since the reduced row echelon form of the matrix involves one free variable the nullspace will have dimension 1. Setting the free variable $\mathbf{x}_4 = 1$ we get

$$x_1 = 10/7, \quad x_2 = 2/7, \quad x_3 = 0$$

Thus $\{(10/7,\ 2/7,\ 0,\ 1)^T\}$ is a basis for the nullspace. The dimension of the column space equals the rank of the matrix which is 3. Thus the column space must be R^3 and we can take as our basis the standard basis $\{\mathbf{e}_1, \mathbf{e}_2, \mathbf{e}_3\}$.

(c) The reduced row echelon form of the matrix is

$$\begin{pmatrix} 1 & 0 & 0 & -0.65 \\ 0 & 1 & 0 & 1.05 \\ 0 & 0 & 1 & 0.75 \end{pmatrix}$$

The set $\{(1,\ 0,\ 0,\ -0.65), (0,\ 1,\ 0,\ 1.05), (0,\ 0,\ 1,\ 0,\ 0.75)\}$ is a basis for the row space. The set $\{(0.65,\ -1.05,\ -0.75,\ 1)^T\}$ is a basis for the nullspace. As in part (b) the column space is R^3 and we can take $\{\mathbf{e}_1, \mathbf{e}_2, \mathbf{e}_3\}$ as our basis.

3 (b) The reduced row echelon form of A is given by

$$U = \begin{pmatrix} 1 & 2 & 0 & 5 & -3 & 0 \\ 0 & 0 & 1 & -1 & 2 & 0 \\ 0 & 0 & 0 & 0 & 0 & 1 \end{pmatrix}$$

The lead variables correspond to columns 1, 3, and 6. Thus $\mathbf{a}_1$, $\mathbf{a}_3$, $\mathbf{a}_6$ form a basis for the column space of A. The remaining column vectors satisfy the following dependency relationships.

$$\begin{aligned} \mathbf{a}_2 &= 2\mathbf{a}_1 \\ \mathbf{a}_4 &= 5\mathbf{a}_1 - \mathbf{a}_3 \\ \mathbf{a}_5 &= -3\mathbf{a}_1 + 2\mathbf{a}_3 \end{aligned}$$

4. (c) consistent, (d) inconsistent, (f) consistent

6. There will be exactly one solution. The condition that $\mathbf{b}$ is in the column space of A guarantees that the system is consistent. If the column vectors are linearly independent, then there is at most one solution. Thus the two conditions together imply exactly one solution.

7. (a) Since $N(A) = \{\mathbf{0}\}$

$$A\mathbf{x} = x_1\mathbf{a}_1 + \cdots + x_n\mathbf{a}_n = \mathbf{0}$$

has only the trivial solution $\mathbf{x} = \mathbf{0}$, and hence $\mathbf{a}_1, \ldots, \mathbf{a}_n$ are linearly independent. The column vectors cannot span R^m since there are only n vectors and $n < m$.

(b) If $\mathbf{b}$ is not in the column space of A, then the system must be inconsistent and hence there will be no solutions. If $\mathbf{b}$ is in the column space of A, then the system will be consistent, so there will be at least one solution. By part (a), the column vectors are linearly independent, so there cannot be more than one solution. Thus, if $\mathbf{b}$ is in the column space of A, then the system will have exactly one solution.

9. (a) If A and B are row equivalent, then they have the same row space and consequently the same rank. Since the dimension of the column space equals the rank it follows that the two column spaces will have the same dimension.

(b) If A and B are row equivalent, then they will have the same row space, however, their column spaces are in general not the same. For example if

$$A = \begin{pmatrix} 1 & 0 \\ 0 & 0 \end{pmatrix} \quad \text{and} \quad B = \begin{pmatrix} 0 & 0 \\ 1 & 0 \end{pmatrix}$$

then A and B are row equivalent but the column space of A is equal to Span($\mathbf{e}_1$) while the column space of B is Span($\mathbf{e}_2$).

10. The column vectors of A and U satisfy the same dependency relations. By inspection one can see that

$$\mathbf{u}_3 = 2\mathbf{u}_1 + \mathbf{u}_2 \quad \text{and} \quad \mathbf{u}_4 = \mathbf{u}_1 + 4\mathbf{u}_2$$

Therefore

$$\mathbf{a}_3 = 2\mathbf{a}_1 + \mathbf{a}_2 = \begin{pmatrix} -6 \\ 10 \\ 4 \\ 2 \end{pmatrix} + \begin{pmatrix} 4 \\ -3 \\ 7 \\ -1 \end{pmatrix} = \begin{pmatrix} -2 \\ 7 \\ 11 \\ 1 \end{pmatrix}$$

and

$$\mathbf{a}_4 = \mathbf{a}_1 + 4\mathbf{a}_2 = \begin{pmatrix} -3 \\ 5 \\ 2 \\ 1 \end{pmatrix} + \begin{pmatrix} 16 \\ -12 \\ 28 \\ -4 \end{pmatrix} = \begin{pmatrix} 13 \\ -7 \\ 30 \\ -3 \end{pmatrix}$$

11. If A is 5×8 with rank 5, then the column space of A will be R^5. So by the Consistency Theorem, the system $A\mathbf{x} = \mathbf{b}$ will be consistent for any $\mathbf{b}$ in R^5. Since A has 8 columns, its reduced row echelon form will involve 3 free variables. A consistent system with free variables must have infinitely many solutions.

12. If U is the reduced row echelon form of A then the given conditions imply that

$$\mathbf{u}_1 = \mathbf{e}_1,\ \mathbf{u}_2 = \mathbf{e}_2,\ \mathbf{u}_3 = \mathbf{u}_1 + 2\mathbf{u}_2,\ \mathbf{u}_4 = \mathbf{e}_3,\ \mathbf{u}_5 = 2\mathbf{u}_1 - \mathbf{u}_2 + 3\mathbf{u}_4$$

Therefore

$$U = \begin{pmatrix} 1 & 0 & 1 & 0 & 2 \\ 0 & 1 & 2 & 0 & -1 \\ 0 & 0 & 0 & 1 & 3 \\ 0 & 0 & 0 & 0 & 0 \end{pmatrix}$$

13. (a) Since A is 5×3 with rank 3, its nullity is 0. Therefore $N(A) = \{\mathbf{0}\}$.

(b) If

$$c_1\mathbf{y}_1 + c_2\mathbf{y}_2 + c_3\mathbf{y}_3 = \mathbf{0}$$

then

$$\begin{aligned} c_1A\mathbf{x}_1 + c_2A\mathbf{x}_2 + c_3A\mathbf{x}_3 &= \mathbf{0} \\ A(c_1\mathbf{x}_1 + c_2\mathbf{x}_2 + c_3\mathbf{x}_3) &= \mathbf{0} \end{aligned}$$

and it follows that $c_1\mathbf{x}_1 + c_2\mathbf{x}_2 + c_3\mathbf{x}_3$ is in $N(A)$. However, we know from part (a) that $N(A) = \{\mathbf{0}\}$. Therefore

$$c_1\mathbf{x}_1 + c_2\mathbf{x}_2 + c_3\mathbf{x}_3 = \mathbf{0}$$

Since $\mathbf{x}_1$, $\mathbf{x}_2$, $\mathbf{x}_3$ are linearly independent it follows that $c_1 = c_2 = c_3 = 0$ and hence $\mathbf{y}_1$, $\mathbf{y}_2$, $\mathbf{y}_3$ are linearly independent.

(c) Since $\dim R^5 = 5$ it takes 5 linearly independent vectors to span the vector space. The vectors $\mathbf{y}_1$, $\mathbf{y}_2$, $\mathbf{y}_3$ do not span R^5 and hence cannot form a basis for R^5.

14. Given A is $m \times n$ with rank n and $\mathbf{y} = A\mathbf{x}$ where $\mathbf{x} \neq \mathbf{0}$. If $\mathbf{y} = \mathbf{0}$ then

$$x_1\mathbf{a}_1 + x_2\mathbf{a}_2 + \cdots + x_n\mathbf{a}_n = \mathbf{0}$$

But this would imply that the columns vectors of A are linearly dependent. Since A has rank n we know that its column vectors must be linearly independent. Therefore $\mathbf{y}$ cannot be equal to $\mathbf{0}$.

15. If the system $A\mathbf{x} = \mathbf{b}$ is consistent, then $\mathbf{b}$ is in the column space of A. Therefore the column space of $(A \mid \mathbf{b})$ will equal the column space of A. Since the rank of a matrix is equal to the dimension of the column space it follows that the rank of $(A \mid \mathbf{b})$ equals the rank of A.

Conversely if $(A \mid \mathbf{b})$ and A have the same rank, then $\mathbf{b}$ must be in the column space of A. If $\mathbf{b}$ were not in the column space of A, then the rank of $(A \mid \mathbf{b})$ would equal $\operatorname{rank}(A) + 1$.

16. (a) If $\mathbf{x} \in N(A)$, then

$$BA\mathbf{x} = B\mathbf{0} = \mathbf{0}$$

and hence $\mathbf{x} \in N(BA)$. Thus $N(A)$ is a subspace of $N(BA)$. On the other hand, if $\mathbf{x} \in N(BA)$, then

$$B(A\mathbf{x}) = BA\mathbf{x} = \mathbf{0}$$

and hence $A\mathbf{x} \in N(B)$. But $N(B) = \{\mathbf{0}\}$ since B is nonsingular. Therefore $A\mathbf{x} = \mathbf{0}$ and hence $\mathbf{x} \in N(A)$. Thus BA and A have the same nullspace. It follows from the Rank-Nullity Theorem that

$$\begin{aligned} \operatorname{rank}(A) &= n - \dim N(A) \\ &= n - \dim N(BA) \\ &= \operatorname{rank}(BA) \end{aligned}$$

(b) By part (a), left multiplication by a nonsingular matrix does not alter the rank. Thus

$$\begin{aligned} \operatorname{rank}(A) = \operatorname{rank}(A^T) &= \operatorname{rank}(C^TA^T) \\ &= \operatorname{rank}((AC)^T) \\ &= \operatorname{rank}(AC) \end{aligned}$$

17. **Corollary 3.6.4.** An $n \times n$ matrix A is nonsingular if and only if the column vectors of A form a basis for R^n.

Proof: It follows from Theorem 3.6.3 that the column vectors of A form a basis for R^n if and only if for each $\mathbf{b} \in R^n$ the system $A\mathbf{x} = \mathbf{b}$ has a unique solution. We claim $A\mathbf{x} = \mathbf{b}$ has a unique solution for each $\mathbf{b} \in R^n$ if and only if A is nonsingular. If A is nonsingular then $\mathbf{x} = A^{-1}\mathbf{b}$ is the unique solution to $A\mathbf{x} = \mathbf{b}$. Conversely, if for each $\mathbf{b} \in R^n$, $A\mathbf{x} = \mathbf{b}$ has a unique solution, then $\mathbf{x} = \mathbf{0}$ is the only solution to $A\mathbf{x} = \mathbf{0}$. Thus it follows from Theorem 1.4.2 that A is nonsingular.

18. If $N(A - B) = R^n$ then the nullity of $A - B$ is n and consequently the rank of $A - B$ must be 0. Therefore

$$\begin{aligned} A - B &= O \\ A &= B \end{aligned}$$

19. **(a)** The column space of B will be a subspace of $N(A)$ if and only if

$$A\mathbf{b}_j = \mathbf{0} \quad \text{for} \quad j = 1, \ldots, n$$

However, the jth column of AB is

$$AB\mathbf{e}_j = A\mathbf{b}_j, \qquad j = 1, \ldots, n$$

Thus the column space of B will be a subspace of $N(A)$ if and only if all the column vectors of AB are $\mathbf{0}$ or equivalently $AB = O$.

(b) Suppose that A has rank r and B has rank k and $AB = O$. By part (a) the column space of B is a subspace of $N(A)$. Since $N(A)$ has dimension $n - r$, it follows that the dimension of the column space of B must be less than or equal to $n - r$. Therefore

$$\text{rank}(A) + \text{rank}(B) = r + k \le r + (n - r) = n$$

20. Let $\mathbf{x}_0$ be a particular solution to $A\mathbf{x} = \mathbf{b}$. If $\mathbf{y} = \mathbf{x}_0 + \mathbf{z}$, where $\mathbf{z} \in N(A)$, then

$$A\mathbf{y} = A\mathbf{x}_0 + A\mathbf{z} = \mathbf{b} + \mathbf{0} = \mathbf{b}$$

and hence $\mathbf{y}$ is also a solution.

Conversely, if $\mathbf{x}_0$ and $\mathbf{y}$ are both solutions to $A\mathbf{x} = \mathbf{b}$ and $\mathbf{z} = \mathbf{y} - \mathbf{x}_0$, then

$$A\mathbf{z} = A\mathbf{y} - A\mathbf{x}_0 = \mathbf{b} - \mathbf{b} = \mathbf{0}$$

and hence $\mathbf{z} \in N(A)$.

21. (a) Since

$$A = \mathbf{x}\mathbf{y}^T = \begin{pmatrix} x_1 \\ x_2 \\ \vdots \\ x_m \end{pmatrix} \mathbf{y}^T = \begin{pmatrix} x_1\mathbf{y}^T \\ x_2\mathbf{y}^T \\ \vdots \\ x_m\mathbf{y}^T \end{pmatrix}$$

the rows of A are all multiples of $\mathbf{y}^T$. Thus $\{\mathbf{y}^T\}$ is a basis for the row space of A. Since

$$A = \mathbf{x}\mathbf{y}^T = \mathbf{x}(y_1, y_2, \ldots, y_n)$$

$$= (y_1\mathbf{x}, y_2\mathbf{x}, \ldots, y_n\mathbf{x})$$

it follows that the columns of A are all multiples of $\mathbf{x}$ and hence $\{\mathbf{x}\}$ is a basis for the column space of A.

(b) Since A has rank 1, the nullity of A is $n - 1$.

22. (a) If $\mathbf{c}$ is a vector in the column space of C, then

$$\mathbf{c} = AB\mathbf{x}$$

for some $\mathbf{x} \in R^r$. Let $\mathbf{y} = B\mathbf{x}$. Since $\mathbf{c} = A\mathbf{y}$, it follows that $\mathbf{c}$ is in the column space of A and hence the column space of C is a subspace of the column space of A.

(b) If $\mathbf{c}^T$ is a row vector of C, then $\mathbf{c}$ is in the column space of C^T. But $C^T = B^TA^T$. Thus, by part (a), $\mathbf{c}$ must be in the column space of B^T and hence $\mathbf{c}^T$ must be in the row space of B.

(c) It follows from part (a) that $\text{rank}(C) \leq \text{rank}(A)$ and it follows from part (b) that $\text{rank}(C) \leq \text{rank}(B)$. Therefore

$$\text{rank}(C) \leq \min\{\text{rank}(A), \text{rank}(B)\}$$

23 (a) In general a matrix E will have linearly independent column vectors if and only if $E\mathbf{x} = \mathbf{0}$ has only the trivial solution $\mathbf{x} = \mathbf{0}$. To show that C has linearly independent column vectors we will show that $C\mathbf{x} \neq \mathbf{0}$ for all $\mathbf{x} \neq \mathbf{0}$ and hence that $C\mathbf{x} = \mathbf{0}$ has only the trivial solution. Let $\mathbf{x}$ be any nonzero vector in R^r and let $\mathbf{y} = B\mathbf{x}$. Since B has linearly independent column vectors it follows that $\mathbf{y} \neq \mathbf{0}$. Similarly since A has linearly independent column vectors, $A\mathbf{y} \neq \mathbf{0}$. Thus

$$C\mathbf{x} = AB\mathbf{x} = A\mathbf{y} \neq \mathbf{0}$$

(b) If A and B both have linearly independent row vectors, then B^T and A^T both have linearly independent column vectors. Since $C^T = B^TA^T$, it follows from part (a) that the column vectors of C^T are linearly independent, and hence the row vectors of C must be linearly independent.

24. (a) If the column vectors of B are linearly dependent then $B\mathbf{x} = \mathbf{0}$ for some nonzero vector $\mathbf{x} \in R^r$. Thus

$$C\mathbf{x} = AB\mathbf{x} = A\mathbf{0} = \mathbf{0}$$

and hence the column vectors of C must be linearly dependent.

(b) If the row vectors of A are linearly dependent then the column vectors of A^T must be linearly dependent. Since $C^T = B^TA^T$, it follows from part (a) that the column vectors of C^T must be linearly dependent. If the column vectors of C^T are linearly dependent, then the row vectors of C must be linearly dependent.

25. (a) Let C denote the right inverse of A and let $\mathbf{b} \in R^m$. If we set $\mathbf{x} = C\mathbf{b}$ then

$$A\mathbf{x} = AC\mathbf{b} = I_m\mathbf{b} = \mathbf{b}$$

Thus if A has a right inverse then $A\mathbf{x} = \mathbf{b}$ will be consistent for each $\mathbf{b} \in R^m$ and consequently the column vectors of A will span R^m.

(b) No set of less than m vectors can span R^m. Thus if $n < m$, then the column vectors of A cannot span R^m and consequently A cannot have a right inverse. If $n \geq m$ then a right inverse is possible.

27. Let B be an $n \times m$ matrix. Since

$$DB = I_m$$

if and only if

$$B^T D^T = I_m^T = I_m$$

it follows that D is a left inverse for B if and only if D^T is a right inverse for B^T.

28. If the column vectors of B are linearly independent, then the row vectors of B^T are linearly independent. Thus B^T has rank m and consequently the column space of B^T is R^m. By Exercise 26, B^T has a right inverse and consequently B must have a left inverse.

29. Let B be an $n \times m$ matrix. If B has a left inverse, then B^T has a right inverse. It follows from Exercise 25 that the column vectors of B^T span R^m. Thus the rank of B^T is m. The rank of B must also be m and consequently the column vectors of B must be linearly independent.

30. Let $\mathbf{u}(1,:), \mathbf{u}(2,:), \ldots, \mathbf{u}(k,:)$ be the nonzero row vectors of U. If

$$c_1\mathbf{u}(1,:) + c_2\mathbf{u}(2,:) + \cdots + c_k\mathbf{u}(k,:) = \mathbf{0}^T$$

then we claim

$$c_1 = c_2 = \cdots = c_k = 0$$

This is true since the leading nonzero entry in $\mathbf{u}(i,:)$ is the only nonzero entry in its column. Let us refer to the column containing the leading nonzero entry of $\mathbf{u}(i,:)$ as $j(i)$. Thus if

$$\mathbf{y}^T = c_1\mathbf{u}(1,:) + c_2\mathbf{u}(2,:) + \cdots + c_k\mathbf{u}(k,:) = \mathbf{0}^T$$

then

$$0 = y_{j(i)} = c_i, \qquad i = 1, \ldots, k$$

and it follows that the nonzero row vectors of U are linearly independent.

MATLAB EXERCISES

1. (a) The column vectors of U will be linearly independent if and only if the rank of U is 4.

(d) The matrices S and T should be inverses.

2. (a) Since

$$r = \text{dim of row space} \leq m$$

and

$$r = \text{dim of column space} \leq n$$

it follows that

$$r \leq \min(m, n)$$

(c) All the rows of A are multiples of $\mathbf{y}^T$ and all of the columns of A are multiples of $\mathbf{x}$. Thus the rank of A is 1.

(d) Since X and Y^T were generated randomly, both should have rank 2 and consequently we would expect that their product should also have rank 2.

3. (a) The column space of C is a subspace of the column space of B. Thus A and B must have the same column space and hence the same rank. Therefore we would expect the rank of A to be 4.

(b) The first four columns of A should be linearly independent and hence should form a basis for the column space of A. The first four columns of the reduced row echelon form of A should be the same as the first four columns of the 8×8 identity matrix. Since the rank is 4, the last four rows should consist entirely of 0's.

(c) If U is the reduced row echelon form of B, then $U = MB$ where M is a product of elementary matrices. If B is an $n \times n$ matrix of rank n, then $U = I$ and $M = B^{-1}$. In this case it follows that the reduced row echelon form of $(B \;\; BX)$ will be

$$B^{-1}(B \;\; BX) = (I \;\; X)$$

If B is $m \times n$ of rank n and $n < m$, then its reduced row echelon form is given by

$$U = MB = \begin{pmatrix} I \\ O \end{pmatrix}$$

It follows that the reduced row echelon form of $(B \;\; BX)$ will be

$$MB(I \;\; X) = \begin{pmatrix} I \\ O \end{pmatrix} (I \;\; X) = \begin{pmatrix} I & X \\ O & O \end{pmatrix}$$

4. (d) The vectors $C\mathbf{y}$ and $\mathbf{b} + c\mathbf{u}$ are equal since

$$C\mathbf{y} = (A + \mathbf{u}\mathbf{v}^T)\mathbf{y} = A\mathbf{y} + c\mathbf{u} = \mathbf{b} + c\mathbf{u}$$

The vectors $C\mathbf{z}$ and $(1 + d)\mathbf{u}$ are equal since

$$C\mathbf{z} = (A + \mathbf{u}\mathbf{v}^T)\mathbf{z} = A\mathbf{z} + d\mathbf{u} = \mathbf{u} + d\mathbf{u}$$

It follows that

$$C\mathbf{x} = C(\mathbf{y} - e\mathbf{z}) = \mathbf{b} + c\mathbf{u} - e(1 + d)\mathbf{u} = \mathbf{b}$$

The rank one update method will fail if $d = -1$. In this case

$$C\mathbf{z} = (1 + d)\mathbf{u} = \mathbf{0}$$

Since $\mathbf{z}$ is nonzero, the matrix C must be singular.

CHAPTER TEST A

1. The statement is true. If S is a subspace of a vector space V, then it is nonempty and it is closed under the operations of V. To show that S, with the operations of addition and scalar multiplication from V, forms a vector space we must show that the eight vector space axioms are satisfied. Since S is closed under scalar multiplication, it follows from Theorem 3.1.1 that if $\mathbf{x}$ is any vector in S, then $\mathbf{0} = 0\mathbf{x}$ is a vector in S and $-1\mathbf{x}$ is the additive inverse of $\mathbf{x}$. So axioms A3 and A4 are satisfied. The remaining six axioms hold for all vectors in V and hence hold for all vectors in S. Thus S is a vector space.
2. The statement is false. The elements of R^3 are 3×1 matrices. Vectors that are in R^2 cannot be in vectors in R^3 since they are only 2×1 matrices.
3. The statement is false. A two dimensional subspace of R^3 corresponds to a plane through the origin in 3-space. If S and T are two different two dimensional subspaces of R^3 then both correspond to planes through the origin and their intersection must correspond to a line through the origin. Thus the intersection cannot consist of just the zero vector.
4. The statement is false in general. See the solution to Exercise 19 of Section 2.
5. The statement is true. See the solution to Exercise 18 of Section 2.
6. The statement is true. See Theorem 3.4.3.
7. The statement is false in general. If $\mathbf{x}_1, \mathbf{x}_2, \ldots, \mathbf{x}_n$ span a vector space V of dimension $k < n$, then they will be linearly dependent since there are more vectors than the dimension of the vector space. For example,

$$\mathbf{x}_1 = \begin{pmatrix} 1 \\ 0 \end{pmatrix}, \quad \mathbf{x}_2 = \begin{pmatrix} 0 \\ 1 \end{pmatrix}, \quad \mathbf{x}_3 = \begin{pmatrix} 1 \\ 1 \end{pmatrix}$$

are vectors that span R^2, but are not linearly independent. Since the dimension of R^2 is 2, any set of more than 2 vectors in R^2 must be linearly dependent.
8. The statement is true. If

$$\text{Span}(\mathbf{x}_1, \mathbf{x}_2, \ldots, \mathbf{x}_k) = \text{Span}(\mathbf{x}_1, \mathbf{x}_2, \ldots, \mathbf{x}_{k-1})$$

then $\mathbf{x}_k$ must be in $\text{Span}(\mathbf{x}_1, \mathbf{x}_2, \ldots, \mathbf{x}_{k-1})$. So $\mathbf{x}_k$ can be written as a linear combination of $\mathbf{x}_1, \mathbf{x}_2, \ldots, \mathbf{x}_{k-1}$ and hence there is a dependency relation among the vectors. Specifically if

$$\mathbf{x}_k = c_1\mathbf{x}_1 + c_2\mathbf{x}_2 + \cdots + c_{k-1}\mathbf{x}_{k-1}$$

then we have the dependency relation

$$c_1\mathbf{x}_1 + c_2\mathbf{x}_2 + \cdots + c_{k-1}\mathbf{x}_{k-1} - 1\mathbf{x}_k = \mathbf{0}$$
9. The statement is true. The rank of A is the dimension of the row space of A. The rank of A^T is the dimension of the row space of A^T. The independent rows of A^T correspond to the independent columns of A. So the rank of A^T equals the dimension of the column space of A. But the row space and

column space of A have the same dimension (Theorem 3.6.5). So A and A^T must have the same rank.

10. If $m \neq n$ then the statement is false since

$$\dim N(A) = n - r \qquad \text{and} \qquad \dim N(A^T) = m - r$$

where r is the rank of A.

CHAPTER TEST B

1. The vectors are linearly dependent since

$$0\mathbf{x}_1 + 0\mathbf{x}_2 + 1\mathbf{x}_3 = 0\mathbf{x}_1 + 0\mathbf{x}_2 + 1\mathbf{0} = \mathbf{0}$$

2. (a) S_1 consists of all vectors of the form

$$\mathbf{x} = \begin{pmatrix} -a \\ a \end{pmatrix}$$

so if

$$\mathbf{x} = \begin{pmatrix} -a \\ a \end{pmatrix} \qquad \text{and} \qquad \mathbf{y} = \begin{pmatrix} -b \\ b \end{pmatrix}$$

are arbitrary vectors in S_1 and c is any scalar then

$$c\mathbf{x} = \begin{pmatrix} -ca \\ ca \end{pmatrix} \in S_1$$

and

$$\mathbf{x} + \mathbf{y} = \begin{pmatrix} -a \\ a \end{pmatrix} + \begin{pmatrix} -b \\ b \end{pmatrix} = \begin{pmatrix} -a-b \\ a+b \end{pmatrix} \in S_1$$

Since S_1 is nonempty and closed under the operations of scalar multiplication and vector addition, it follow that S_1 is a subspace of R^2.

(b) S_2 is not a subspace of R^2 since it is not closed under addition. The vectors

$$\mathbf{x} = \begin{pmatrix} 1 \\ 0 \end{pmatrix} \qquad \text{and} \qquad \mathbf{y} = \begin{pmatrix} 0 \\ 1 \end{pmatrix}$$

are both in S_2, however,

$$\mathbf{x} + \mathbf{y} = \begin{pmatrix} 1 \\ 1 \end{pmatrix}$$

is not in S_2.

3. (a)

$$\left(\begin{array}{ccccc|c} 1 & 3 & 1 & 3 & 4 & 0 \\ 0 & 0 & 1 & 1 & 1 & 0 \\ 0 & 0 & 2 & 2 & 2 & 0 \\ 0 & 0 & 3 & 3 & 3 & 0 \end{array}\right) \to \left(\begin{array}{ccccc|c} 1 & 3 & 0 & 2 & 3 & 0 \\ 0 & 0 & 1 & 1 & 1 & 0 \\ 0 & 0 & 0 & 0 & 0 & 0 \\ 0 & 0 & 0 & 0 & 0 & 0 \end{array}\right)$$

The free variables are x_2, x_4, and x_5. If we set $x_2 = a$, $x_4 = b$, and $x_5 = c$, then

$$x_1 = -3a - 2b - 3c \qquad \text{and} \qquad x_3 = -b - c$$

Thus $N(A)$ consists of all vectors of the form

$$\mathbf{x} = \begin{pmatrix} -3a-2b-3c \\ a \\ -b-c \\ b \\ c \end{pmatrix} = a\begin{pmatrix} -3 \\ 1 \\ 0 \\ 0 \\ 0 \end{pmatrix} + b\begin{pmatrix} -2 \\ 0 \\ -1 \\ 1 \\ 0 \end{pmatrix} + c\begin{pmatrix} -3 \\ 0 \\ -1 \\ 0 \\ 1 \end{pmatrix}$$

The vectors

$$\mathbf{x}_1 = \begin{pmatrix} -3 \\ 1 \\ 0 \\ 0 \\ 0 \end{pmatrix}, \quad \mathbf{x}_2 = \begin{pmatrix} -2 \\ 0 \\ -1 \\ 1 \\ 0 \end{pmatrix}, \quad \mathbf{x}_3 = \begin{pmatrix} -3 \\ 0 \\ -1 \\ 0 \\ 1 \end{pmatrix}$$

form a basis for $N(A)$.

(b) The lead 1's occur in the first and third columns of the echelon form. Therefore

$$\mathbf{a}_1 = \begin{pmatrix} 1 \\ 0 \\ 0 \\ 0 \end{pmatrix}, \quad \mathbf{a}_3 = \begin{pmatrix} 1 \\ 1 \\ 2 \\ 3 \end{pmatrix}$$

form a basis for the column space of A.

4. The columns of the matrix that correspond to the lead variables are linearly independent and span the column space of the matrix. So the dimension of the column space is equal to the number of lead variables in any row echelon form of the matrix. If there are r lead variables then there are $n-r$ free variables. By the Rank-Nullity Theorem the dimension of the nullspace is $n-r$. So the dimension of the nullspace is equal to the number of free variables in any echelon form of the matrix.

5. (a) One dimensional subspaces correspond to lines through the origin in 3-space. If the first subspace U_1 is the span of a vector $\mathbf{u}_1$ and the second subspace U_2 is the span of a vector $\mathbf{u}_2$ and the vectors $\mathbf{u}_1$ and $\mathbf{u}_2$ are linearly independent, then the two lines will only intersect at the origin and consequently we will have $U_1 \cap U_2 = \{\mathbf{0}\}$.

(b) Two dimensional subspaces correspond to planes through the origin in 3-space. Any two distinct planes through the origin will intersect in a line. So $V_1 \cap V_2$ must contain infinitely many vectors.

6. (a) If

$$A = \begin{pmatrix} a & b \\ b & c \end{pmatrix}, \quad B = \begin{pmatrix} d & e \\ e & f \end{pmatrix}$$

are arbitrary symmetric matrices and α is any scalar, then

$$\alpha A = \begin{pmatrix} \alpha a & \alpha b \\ \alpha b & \alpha c \end{pmatrix} \quad \text{and} \quad A + B = \begin{pmatrix} a+d & b+e \\ b+e & c+f \end{pmatrix}$$

are both symmetric. Therefore S is closed under the operations of scalar multiplication and vector addition and hence S is a subspace of $R^{2\times 2}$.

(b) The vectors

$$E_1 = \begin{pmatrix} 1 & 0 \\ 0 & 0 \end{pmatrix}, \quad E_2 = \begin{pmatrix} 0 & 1 \\ 1 & 0 \end{pmatrix}, \quad E_3 = \begin{pmatrix} 0 & 0 \\ 0 & 1 \end{pmatrix}$$

are linearly independent and they span S. Therefore they form a basis for S.

7. (a) If A is 6×4 with rank 4, then by the Rank-Nullity Theorem $\dim N(A) = 0$ and consequently $N(A) = \{\mathbf{0}\}$.

(b) The column vectors of A are linearly independent since the rank of A is 4, however, they do not span R^6 since you need 6 linearly independent vectors to spanR^6.

(c) By the Consistency Theorem if $\mathbf{b}$ is in the column space of A then the system is consistent. The condition that the column vectors of A are linearly independent implies that there cannot be more than 1 solution. Therefore there must be exactly 1 solution.

8. (a) The dimension of R^3 is 3, so any collection of more than 3 vectors must be linearly dependent.

(b) Since $\dim R^3 = 3$, it takes 3 linearly independent vectors to span R^3. No 2 vectors can span, so $\mathbf{x}_1$ and $\mathbf{x}_2$ do not span R^3.

(c) The matrix

$$X = (\mathbf{x}_1, \mathbf{x}_2, \mathbf{x}_3) = \begin{pmatrix} 1 & 1 & 1 \\ 2 & 3 & 5 \\ 2 & 3 & 5 \end{pmatrix}$$

only has 2 linearly independent row vectors, so the dimension of the rowspace and dimension of the column space both must be equal to 2. Therefore $\mathbf{x}_1$, $\mathbf{x}_2$, $\mathbf{x}_3$ are linearly dependent and only span a 2-dimensional subspace of R^3. The vectors to not form a basis for R^3 since they are linearly dependent.

(d) If we set $A = (\mathbf{x}_1, \mathbf{x}_2, \mathbf{x}_4)$, then

$$\det(A) = \begin{vmatrix} 1 & 1 & 1 \\ 2 & 3 & 2 \\ 2 & 3 & 3 \end{vmatrix} = \begin{vmatrix} 1 & 1 & 1 \\ 0 & 1 & 0 \\ 0 & 1 & 1 \end{vmatrix} = 1$$

Therefore $\mathbf{x}_1$, $\mathbf{x}_2$, $\mathbf{x}_3$ are linearly independent. Since $\dim R^3 = 3$, the three vectors will span and form a basis for R^3.

9. If

$$c_1\mathbf{y}_1 + c_2\mathbf{y}_2 + c_3\mathbf{y}_3 = \mathbf{0}$$

then

$$c_1A\mathbf{x}_1 + c_2A\mathbf{x}_2 + c_3A\mathbf{x}_3 = A\mathbf{0} = \mathbf{0}$$

Multiplying through by A^{-1} we get

$$c_1\mathbf{x}_1 + c_2\mathbf{x}_2 + c_3\mathbf{x}_3 = \mathbf{0}$$

Since $\mathbf{x}_1$, $\mathbf{x}_2$, $\mathbf{x}_3$ are linearly independent, it follows that $c_1 = c_2 = c_3 = 0$. Therefore $\mathbf{y}_1$, $\mathbf{y}_2$, $\mathbf{y}_3$ are linearly independent.

10. **(a)** The rank of A equals the dimension of the column space of A which is 3. By the Rank-Nullity Theorem, $\dim N(A) = 5 - 3 = 2$.

(b) Since $\mathbf{a}_1$, $\mathbf{a}_2$, $\mathbf{a}_3$ are linearly independent, the first three columns of the reduced row echelon form U will be

$$\mathbf{u}_1 = \mathbf{e}_1, \quad \mathbf{u}_2 = \mathbf{e}_2, \quad \mathbf{u}_3 = \mathbf{e}_3$$

The remaining columns of U satisfy the same dependency relations that the column vectors of A satisfy. Therefore

$$\begin{aligned} \mathbf{u}_4 &= \mathbf{u}_1 + 3\mathbf{u}_2 + \mathbf{u}_3 = \mathbf{e}_1 + 3\mathbf{e}_2 + \mathbf{e}_3 \\ \mathbf{u}_5 &= 2\mathbf{u}_1 - \mathbf{u}_3 = 2\mathbf{e}_1 - \mathbf{e}_3 \end{aligned}$$

and it follows that

$$U = \begin{pmatrix} 1 & 0 & 0 & 1 & 2 \\ 0 & 1 & 0 & 3 & 0 \\ 0 & 0 & 1 & 1 & -1 \\ 0 & 0 & 0 & 0 & 0 \\ 0 & 0 & 0 & 0 & 0 \\ 0 & 0 & 0 & 0 & 0 \end{pmatrix}$$

11. **(a)** If $U = (\mathbf{u}_1, \mathbf{u}_2)$, then the transition matrix corresponding to a change of basis from $[\mathbf{e}_1, \mathbf{e}_2]$ to $[\mathbf{u}_1, \mathbf{u}_2]$ is

$$U^{-1} = \begin{pmatrix} 7 & -2 \\ -3 & 1 \end{pmatrix}$$

(b) Let $V = (\mathbf{v}_1, \mathbf{v}_2)$. If $\mathbf{x} = V\mathbf{d} = U\mathbf{c}$ then $\mathbf{c} = U^{-1}V\mathbf{d}$ and hence the transition matrix corresponding to a change of basis from $[\mathbf{v}_1, \mathbf{v}_2]$ to $[\mathbf{u}_1, \mathbf{u}_2]$ is

$$U^{-1}V = \begin{pmatrix} 7 & -2 \\ -3 & 1 \end{pmatrix} \begin{pmatrix} 5 & 4 \\ 2 & 9 \end{pmatrix} = \begin{pmatrix} 31 & 10 \\ -13 & -3 \end{pmatrix}$$

Chapter 4

SECTION 1

2. $x_1 = r\cos\theta$, $x_2 = r\sin\theta$ where $r = (x_1^2 + x_2^2)^{1/2}$ and θ is the angle between $\mathbf{x}$ and $\mathbf{e}_1$.

$$\begin{aligned} L(\mathbf{x}) &= (r\cos\theta\cos\alpha - r\sin\theta\sin\alpha, r\cos\theta\sin\alpha + r\sin\theta\cos\alpha)^T \\ &= (r\cos(\theta+\alpha), r\sin(\theta+\alpha))^T \end{aligned}$$

The linear transformation L has the effect of rotating a vector by an α in the counterclockwise direction.

3. If $\alpha \neq 1$ then

$$L(\alpha\mathbf{x}) = \alpha\mathbf{x} + \mathbf{a} \neq \alpha\mathbf{x} + \alpha\mathbf{a} = \alpha L(\mathbf{x})$$

The addition property also fails

$$\begin{aligned} L(\mathbf{x}+\mathbf{y}) &= \mathbf{x}+\mathbf{y}+\mathbf{a} \\ L(\mathbf{x}) + L(\mathbf{y}) &= \mathbf{x}+\mathbf{y}+2\mathbf{a} \end{aligned}$$

4. Let

$$\mathbf{u}_1 = \begin{pmatrix} 1 \\ 2 \end{pmatrix}, \quad \mathbf{u}_2 = \begin{pmatrix} 1 \\ -1 \end{pmatrix}, \quad \mathbf{x} = \begin{pmatrix} 7 \\ 5 \end{pmatrix}$$

To determine $L(\mathbf{x})$ we must first express $\mathbf{x}$ as a linear combination

$$\mathbf{x} = c_1\mathbf{u}_1 + c_2\mathbf{u}_2$$

To do this we must solve the system $U\mathbf{c} = \mathbf{x}$ for $\mathbf{c}$. The solution is $\mathbf{c} = (4,3)^T$ and it follows that

$$L(\mathbf{x}) = L(4\mathbf{u}_1 + 3\mathbf{u}_2) = 4L(\mathbf{u}_1) + 3L(\mathbf{u}_2) = 4\begin{pmatrix} -2 \\ 3 \end{pmatrix} + 3\begin{pmatrix} 5 \\ 2 \end{pmatrix} = \begin{pmatrix} 7 \\ 18 \end{pmatrix}$$

8. **(a)**

$$L(\alpha A) = C(\alpha A) + (\alpha A)C = \alpha(CA + AC) = \alpha L(A)$$

and

$$\begin{aligned} L(A+B) &= C(A+B) + (A+B)C = CA + CB + AC + BC \\ &= (CA + AC) + (CB + BC) = L(A) + L(B) \end{aligned}$$

Therefore L is a linear operator.

(b) $L(\alpha A + \beta B) = C^2(\alpha A + \beta B) = \alpha C^2 A + \beta C^2 B = \alpha L(A) + \beta L(B)$
Therefore L is a linear operator.

(c) If $C \neq O$ then L is not a linear operator. For example,

$$L(2I) = (2I)^2 C = 4C \neq 2C = 2L(I)$$

10. If $f, g \in C[0,1]$ then

$$\begin{aligned} L(\alpha f + \beta g) &= \int_0^x (\alpha f(t) + \beta g(t))dt \\ &= \alpha \int_0^x f(t)dt + \beta \int_0^x g(t)dt \\ &= \alpha L(f) + \beta L(g) \end{aligned}$$

Thus L is a linear transformation from $C[0,1]$ to $C[0,1]$.

12. If L is a linear operator from V into W use mathematical induction to prove

$$L(\alpha_1 \mathbf{v}_1 + \alpha_2 \mathbf{v}_2 + \cdots + \alpha_n \mathbf{v}_n) = \alpha_1 L(\mathbf{v}_1) + \alpha_2 L(\mathbf{v}_2) + \cdots + \alpha_n L(\mathbf{v}_n).$$

Proof: In the case $n = 1$

$$L(\alpha_1 \mathbf{v}_1) = \alpha_1 L(\mathbf{v}_1)$$

Let us assume the result is true for any linear combination of k vectors and apply L to a linear combination of $k+1$ vectors.

$$\begin{aligned} L(\alpha_1 \mathbf{v}_1 + \cdots + \alpha_k \mathbf{v}_k + \alpha_{k+1}\mathbf{v}_{k+1}) &= L([\alpha_1 \mathbf{v}_1 + \cdots + \alpha_k \mathbf{v}_k] + [\alpha_{k+1}\mathbf{v}_{k+1}]) \\ &= L(\alpha_1 \mathbf{v}_1 + \cdots + \alpha_k \mathbf{v}_k) + L(\alpha_{k+1}\mathbf{v}_{k+1}) \\ &= \alpha_1 L(\mathbf{v}_1) + \cdots + \alpha_k L(\mathbf{v}_k) + \alpha_{k+1} L(\mathbf{v}_{k+1}) \end{aligned}$$

The result follows then by mathematical induction.

13. If $\mathbf{v}$ is any element of V then

$$\mathbf{v} = \alpha_1 \mathbf{v}_1 + \alpha_2 \mathbf{v}_2 + \cdots + \alpha_n \mathbf{v}_n$$

Since $L_1(\mathbf{v}_i) = L_2(\mathbf{v}_i)$ for $i = 1, \ldots, n$, it follows that

$$\begin{aligned} L_1(\mathbf{v}) &= \alpha_1 L_1(\mathbf{v}_1) + \alpha_2 L_1(\mathbf{v}_2) + \cdots + \alpha_n L_1(\mathbf{v}_n) \\ &= \alpha_1 L_2(\mathbf{v}_1) + \alpha_2 L_2(\mathbf{v}_2) + \cdots + \alpha_n L_2(\mathbf{v}_n) \\ &= L_2(\alpha_1 \mathbf{v}_1 + \cdots + \alpha_n \mathbf{v}_n) \\ &= L_2(\mathbf{v}) \end{aligned}$$

14. Let L be a linear transformation from R^1 to R^1. If $L(1) = \mathbf{a}$ then

$$L(\mathbf{x}) = L(x\mathbf{1}) = xL(\mathbf{1}) = x\mathbf{a} = a\mathbf{x}$$

15. The proof is by induction on n. In the case $n = 1$, L^1 is a linear operator since $L^1 = L$. We will show that if L^m is a linear operator on V then L^{m+1} is also a linear operator on V. This follows since

$$L^{m+1}(\alpha\mathbf{v}) = L(L^m(\alpha\mathbf{v})) = L(\alpha L^m(\mathbf{v})) = \alpha L(L^m(\mathbf{v})) = \alpha L^{m+1}(\mathbf{v})$$

and

$$\begin{aligned} L^{m+1}(\mathbf{v}_1 + \mathbf{v}_2) &= L(L^m(\mathbf{v}_1 + \mathbf{v}_2)) \\ &= L(L^m(\mathbf{v}_1) + L^m(\mathbf{v}_2)) \\ &= L(L^m(\mathbf{v}_1)) + L(L^m(\mathbf{v}_2)) \\ &= L^{m+1}(\mathbf{v}_1) + L^{m+1}(\mathbf{v}_2) \end{aligned}$$

16. If $\mathbf{v}_1, \mathbf{v}_2 \in V$, then

$$\begin{aligned} L(\alpha\mathbf{v}_1 + \beta\mathbf{v}_2) &= L_2(L_1(\alpha\mathbf{v}_1 + \beta\mathbf{v}_2)) \\ &= L_2(\alpha L_1(\mathbf{v}) + \beta L_1(\mathbf{v}_2)) \\ &= \alpha L_2(L_1(\mathbf{v}_1)) + \beta L_2(L_1(\mathbf{v}_2)) \\ &= \alpha L(\mathbf{v}_1) + \beta L(\mathbf{v}_2) \end{aligned}$$

Therefore L is a linear transformation.

17. (b) $\ker(L) = \text{Span}(\mathbf{e}_3)$, $L(R^3) = \text{Span}(\mathbf{e}_1, \mathbf{e}_2)$

18. (c) $L(S) = \text{Span}((1, 1, 1)^T)$

19. (b) If $p(x) = ax^2 + bx + c$ is in $\ker(L)$, then

$$L(p) = (ax^2 + bx + c) - (2ax + b) = ax^2 + (b - 2a)x + (c - b)$$

must equal the zero polynomial $z(x) = 0x^2 + 0x + 0$. Equating coefficients we see that $a = b = c = 0$ and hence $\ker(L) = \{0\}$. The range of L is all of P_3. To see this note that if $p(x) = ax^2 + bx + c$ is any vector in P_3 and we define $q(x) = ax^2 + (b + 2a)x + c + b + 2a$ then

$$L(q(x)) = (ax^2 + (b+2a)x + c + b + 2a) - (2ax + b + 2a) = ax^2 + bx + c = p(x)$$

20. If $\mathbf{0}_V$ denotes the zero vector in V and $\mathbf{0}_W$ is the zero vector in W then $L(\mathbf{0}_V) = \mathbf{0}_W$. Since $\mathbf{0}_W$ is in T, it follows that $\mathbf{0}_V$ is in $L^{-1}(T)$ and hence $L^{-1}(T)$ is nonempty. If $\mathbf{v}$ is in $L^{-1}(T)$, then $L(\mathbf{v}) \in T$. It follows that $L(\alpha\mathbf{v}) = \alpha L(\mathbf{v})$ is in T and hence $\alpha\mathbf{v} \in L^{-1}(T)$. If $\mathbf{v}_1, \mathbf{v}_2 \in L^{-1}(T)$, then $L(\mathbf{v}_1)$, $L(\mathbf{v}_2)$ are in T and hence

$$L(\mathbf{v}_1 + \mathbf{v}_2) = L(\mathbf{v}_1) + L(\mathbf{v}_2)$$

is also an element of $L(T)$. Thus $\mathbf{v}_1 + \mathbf{v}_2 \in L^{-1}(T)$ and therefore $L^{-1}(T)$ is a subspace of V.

21. Suppose L is one-to-one and $\mathbf{v} \in \ker(L)$.

$$L(\mathbf{v}) = \mathbf{0}_W \quad \text{and} \quad L(\mathbf{0}_V) = \mathbf{0}_W$$

Since L is one-to-one, it follows that $\mathbf{v} = \mathbf{0}_V$. Therefore $\ker(L) = \{\mathbf{0}_V\}$.

Conversely, suppose $\ker(L) = \{\mathbf{0}_V\}$ and $L(\mathbf{v}_1) = L(\mathbf{v}_2)$. Then

$$L(\mathbf{v}_1 - \mathbf{v}_2) = L(\mathbf{v}_1) - L(\mathbf{v}_2) = \mathbf{0}_W$$

Therefore $\mathbf{v}_1 - \mathbf{v}_2 \in \ker(L)$ and hence

$$\begin{aligned} \mathbf{v}_1 - \mathbf{v}_2 &= \mathbf{0}_V \\ \mathbf{v}_1 &= \mathbf{v}_2 \end{aligned}$$

So L is one-to-one.

22. To show that L maps R^3 onto R^3 we must show that for any vector $\mathbf{y} \in R^3$ there exists a vector $\mathbf{x} \in R^3$ such that $L(\mathbf{x}) = \mathbf{y}$. This is equivalent to showing that the linear system

$$\begin{array}{lllllll} x_1 & & & & & = & y_1 \\ x_1 & + & x_2 & & & = & y_2 \\ x_1 & + & x_2 & + & x_3 & = & y_3 \end{array}$$

is consistent. This system is consistent since the coefficient matrix is nonsingular.

24. (a) $L(R^2) = \{A\mathbf{x} \mid \mathbf{x} \in R^2\}$
$= \{x_1\mathbf{a}_1 + x_2\mathbf{a}_2 \mid x_1, x_2 \text{ real }\}$
$=$ the column space of A

(b) If A is nonsingular, then A has rank 2 and it follows that its column space must be R^2. By part (a), $L(R^2) = R^2$.

25. (a) If $p = ax^2 + bx + c \in P_3$, then

$$D(p) = 2ax + b$$

Thus

$$D(P_3) = \text{Span}(1, x) = P_2$$

The operator is not one-to-one, for if $p_1(x) = ax^2 + bx + c_1$ and $p_2(x) = ax^2 + bx + c_2$ where $c_2 \neq c_1$, then $D(p_1) = D(p_2)$.

(b) The subspace S consists of all polynomials of the form $ax^2 + bx$. If $p_1 = a_1x^2 + b_1x$, $p_2 = a_2x^2 + b_2x$ and $D(p_1) = D(p_2)$, then

$$2a_1x + b_1 = 2a_2x + b_2$$

and it follows that $a_1 = a_2$, $b_1 = b_2$. Thus $p_1 = p_2$ and hence D is one-to-one. D does not map S onto P_3 since $D(S) = P_2$.

SECTION 2

7. (a) $\mathcal{I}(\mathbf{e}_1) = 0\mathbf{y}_1 + 0\mathbf{y}_2 + 1\mathbf{y}_3$
$\mathcal{I}(\mathbf{e}_2) = 0\mathbf{y}_1 + 1\mathbf{y}_2 - 1\mathbf{y}_3$
$\mathcal{I}(\mathbf{e}_3) = 1\mathbf{y}_1 - 1\mathbf{y}_2 + 0\mathbf{y}_3$

10. (c) $\begin{pmatrix} \frac{1}{3} & 0 & 0 \\ 0 & \frac{1}{3} & 0 \\ 0 & 0 & 1 \end{pmatrix}$

11. (a) $YP = \begin{pmatrix} 0 & 1 & 0 \\ -1 & 0 & 0 \\ 0 & 0 & 1 \end{pmatrix} \begin{pmatrix} 0 & 0 & -1 \\ 0 & 1 & 0 \\ 1 & 0 & 0 \end{pmatrix} = \begin{pmatrix} 0 & 1 & 0 \\ 0 & 0 & 1 \\ 1 & 0 & 0 \end{pmatrix}$

(b) $PY = \begin{pmatrix} 0 & 0 & -1 \\ 0 & 1 & 0 \\ 1 & 0 & 0 \end{pmatrix} \begin{pmatrix} 0 & 1 & 0 \\ -1 & 0 & 0 \\ 0 & 0 & 1 \end{pmatrix} = \begin{pmatrix} 0 & 0 & -1 \\ -1 & 0 & 0 \\ 0 & 1 & 0 \end{pmatrix}$

(c) $PR = \begin{pmatrix} \frac{1}{\sqrt{2}} & 0 & -\frac{1}{\sqrt{2}} \\ 0 & 1 & 0 \\ \frac{1}{\sqrt{2}} & 0 & \frac{1}{\sqrt{2}} \end{pmatrix} \begin{pmatrix} 1 & 0 & 0 \\ 0 & 0 & 1 \\ 0 & -1 & 0 \end{pmatrix} = \begin{pmatrix} \frac{1}{\sqrt{2}} & \frac{1}{\sqrt{2}} & 0 \\ 0 & 0 & 1 \\ \frac{1}{\sqrt{2}} & -\frac{1}{\sqrt{2}} & 0 \end{pmatrix}$

(d) $RP = \begin{pmatrix} 1 & 0 & 0 \\ 0 & 0 & 1 \\ 0 & -1 & 0 \end{pmatrix} \begin{pmatrix} \frac{1}{\sqrt{2}} & 0 & -\frac{1}{\sqrt{2}} \\ 0 & 1 & 0 \\ \frac{1}{\sqrt{2}} & 0 & \frac{1}{\sqrt{2}} \end{pmatrix} = \begin{pmatrix} \frac{1}{\sqrt{2}} & 0 & -\frac{1}{\sqrt{2}} \\ \frac{1}{\sqrt{2}} & 0 & \frac{1}{\sqrt{2}} \\ 0 & -1 & 0 \end{pmatrix}$

(e)

$$YPR = \begin{pmatrix} \frac{1}{\sqrt{2}} & \frac{1}{\sqrt{2}} & 0 \\ -\frac{1}{\sqrt{2}} & \frac{1}{\sqrt{2}} & 0 \\ 0 & 0 & 1 \end{pmatrix} \begin{pmatrix} 0 & 0 & 1 \\ 0 & 1 & 0 \\ -1 & 0 & 0 \end{pmatrix} \begin{pmatrix} 1 & 0 & 0 \\ 0 & \frac{1}{\sqrt{2}} & \frac{1}{\sqrt{2}} \\ 0 & -\frac{1}{\sqrt{2}} & \frac{1}{\sqrt{2}} \end{pmatrix}$$

$$= \begin{pmatrix} 0 & 0 & 1 \\ 0 & 1 & 0 \\ -1 & 0 & 0 \end{pmatrix}$$

(f)

$$RPY = \begin{pmatrix} 1 & 0 & 0 \\ 0 & \frac{1}{\sqrt{2}} & \frac{1}{\sqrt{2}} \\ 0 & -\frac{1}{\sqrt{2}} & \frac{1}{\sqrt{2}} \end{pmatrix} \begin{pmatrix} 0 & 0 & 1 \\ 0 & 1 & 0 \\ -1 & 0 & 0 \end{pmatrix} \begin{pmatrix} \frac{1}{\sqrt{2}} & \frac{1}{\sqrt{2}} & 0 \\ -\frac{1}{\sqrt{2}} & \frac{1}{\sqrt{2}} & 0 \\ 0 & 0 & 1 \end{pmatrix}$$

$$= \begin{pmatrix} 0 & 0 & 1 \\ -1 & 0 & 0 \\ 0 & -1 & 0 \end{pmatrix}$$

12. (a) If Y is the yaw matrix and we expand $\det(Y)$ along its third row we get

$$\det(Y) = \cos^2 u + \sin^2 u = 1$$

Similarly, if we expand the determinant pitch matrix P along its second and expand the determinant of the roll matrix R along its first row we get

$$\begin{aligned} \det(P) &= \cos^2 v + \sin^2 v = 1 \\ \det(R) &= \cos^2 w + \sin^2 w = 1 \end{aligned}$$

(b) If Y is a yaw matrix with yaw angle u then

$$Y^T = \begin{pmatrix} \cos u & -\sin u & 0 \\ \sin u & \cos u & 0 \\ 0 & 0 & 1 \end{pmatrix} = \begin{pmatrix} \cos(-u) & \sin(-u) & 0 \\ -\sin(-u) & \cos(-u) & 0 \\ 0 & 0 & 1 \end{pmatrix}$$

so Y^T is the matrix representing a yaw transformation with angle $-u$. It is easily verified that $Y^TY = I$ and hence that $Y^{-1} = Y^T$.

(c) By the same reasoning used in part (b) you can show that for the pitch matrix P and roll matrix R their inverses are their transposes. So if $Q = YPR$ then Q is nonsingular and

$$Q^{-1} = (YPR)^{-1} = R^{-1}P^{-1}Y^{-1} = R^TP^TY^T$$

14. (b) $\begin{pmatrix} 3/2 \\ -2 \end{pmatrix}$; (c) $\begin{pmatrix} 3/2 \\ 0 \end{pmatrix}$

16. If $L(\mathbf{x}) = \mathbf{0}$ for some $\mathbf{x} \neq \mathbf{0}$ and A is the standard matrix representation of L, then $A\mathbf{x} = \mathbf{0}$. It follows from Theorem 1.4.2 that A is singular.

17. The proof is by induction on m. In the case that $m = 1$, $A^1 = A$ represents $L^1 = L$. If now A^k is the matrix representing L^k and if $\mathbf{x}$ is the coordinate vector of $\mathbf{v}$, then $A^k\mathbf{x}$ is the coordinate vector of $L^k(\mathbf{v})$. Since

$$L^{k+1}(\mathbf{v}) = L(L^k(\mathbf{v}))$$

it follows that

$$AA^k\mathbf{x} = A^{k+1}\mathbf{x}$$

is the coordinate vector of $L^{k+1}(\mathbf{v})$.

18. (b) $\begin{pmatrix} -5 & -2 & 4 \\ 3 & 2 & -2 \end{pmatrix}$

19. If $\mathbf{x} = [\mathbf{v}]_E$, then $A\mathbf{x} = [L_1(\mathbf{v})]_F$ and $B(A\mathbf{x}) = [L_2(L_1(\mathbf{v}))]_G$. Thus, for all $\mathbf{v} \in V$

$$(BA)[\mathbf{v}]_E = [L_2 \circ L_1(\mathbf{v})]_G$$

Hence BA is the matrix representing $L_2 \circ L_1$ with respect to E and G.

20. (a) Since A is the matrix representing L with respect to E and F, it follows that $L(\mathbf{v}) = \mathbf{0}_W$ if and only if $A[\mathbf{v}]_E = \mathbf{0}$. Thus $\mathbf{v} \in \ker(L)$ if and only if $[\mathbf{v}]_E \in N(A)$.

(b) Since A is the matrix representing L with respect to E and F, then it follows that $\mathbf{w} = L(\mathbf{v})$ if and only if $[\mathbf{w}]_F = A[\mathbf{v}]_E$. Thus, $\mathbf{w} \in L(V)$ if and only if $[\mathbf{w}]_F$ is in the column space of A.

SECTION 3

7. If A is similar to B then there exists a nonsingular matrix S_1 such that $A = S_1^{-1}BS_1$. Since B is similar to C there exists a nonsingular matrix S_2 such that $B = S_2^{-1}CS_2$. It follows that

$$A = S_1^{-1}BS_1 = S_1^{-1}S_2^{-1}CS_2S_1$$

If we set $S = S_2 S_1$, then S is nonsingular and $S^{-1} = S_1^{-1} S_2^{-1}$. Thus $A = S^{-1}CS$ and hence A is similar to C.

8. (a) If $A = S\Lambda S^{-1}$, then $AS = \Lambda S$. If $\mathbf{s}_i$ is the ith column of S then $A\mathbf{s}_i$ is the ith column of AS and $\lambda_i \mathbf{s}_i$ is the ith column of ΛS. Thus

$$A\mathbf{s}_i = \lambda_i \mathbf{s}_i, \qquad i = 1, \ldots, n$$

(b) The proof is by induction on k. In the case $k = 1$ we have by part (a):

$$A\mathbf{x} = \alpha_1 A\mathbf{s}_1 + \cdots + \alpha_n A\mathbf{s}_n = \alpha_1 \lambda_1 \mathbf{s}_1 + \cdots + \alpha_n \lambda_n \mathbf{s}_n$$

If the result holds in the case $k = m$

$$A^m \mathbf{x} = \alpha_1 \lambda_1^m \mathbf{s}_1 + \cdots + \alpha_n \lambda_n^m \mathbf{s}_n$$

then

$$\begin{aligned} A^{m+1}\mathbf{x} &= \alpha_1 \lambda_1^m A\mathbf{s}_1 + \cdots + \alpha_n \lambda_n^m A\mathbf{s}_n \\ &= \alpha_1 \lambda_1^{m+1} \mathbf{s}_1 + \cdots + \alpha_n \lambda_n^{m+1} \mathbf{s}_n \end{aligned}$$

Therefore by mathematical induction the result holds for all natural numbers k.

(c) If $|\lambda_i| < 1$ then $\lambda_i^k \to 0$ as $k \to \infty$. It follows from part (b) that $A^k \mathbf{x} \to \mathbf{0}$ as $k \to \infty$.

9. If $A = ST$ then

$$S^{-1}AS = S^{-1}STS = TS = B$$

Therefore B is similar to A.

10. If A and B are similar, then there exists a nonsingular matrix S such that

$$A = SBS^{-1}$$

If we set

$$T = BS^{-1}$$

then

$$A = ST \qquad \text{and} \qquad B = TS$$

11. If $B = S^{-1}AS$, then

$$\begin{aligned} \det(B) &= \det(S^{-1}AS) \\ &= \det(S^{-1})\det(A)\det(S) \\ &= \det(A) \end{aligned}$$

since

$$\det(S^{-1}) = \frac{1}{\det(S)}$$

12. (a) If $B = S^{-1}AS$, then

$$\begin{aligned} B^T &= (S^{-1}AS)^T \\ &= S^T A^T (S^{-1})^T \\ &= S^T A^T (S^T)^{-1} \end{aligned}$$

Therefore B^T is similar to A^T.

(b) If $B = S^{-1}AS$, then one can prove using mathematical induction that

$$B^k = S^{-1}A^kS$$

for any positive integer k. Therefore that B^k and A^k are similar for any positive integer k.

13. If A is similar to B and A is nonsingular, then

$$A = SBS^{-1}$$

and hence

$$B = S^{-1}AS$$

Since B is a product of nonsingular matrices it is nonsingular and

$$B^{-1} = (S^{-1}AS)^{-1} = S^{-1}A^{-1}S$$

Therefore B^{-1} and A^{-1} are similar.

14. If A and B are similar, then there exists a nonsingular matrix S such that $B = SAS^{-1}$.

(a) $A - \lambda I$ and $B - \lambda I$ are similar since

$$S(A - \lambda I)S^{-1} = SAS^{-1} - \lambda SIS^{-1} = B - \lambda I$$

(b) Since $A - \lambda I$ and $B - \lambda I$ are similar, it follows from Exercise 11 that their determinants are equal.

15. (a) Let $C = AB$ and $E = BA$. The diagonal entries of C and E are given by

$$c_{ii} = \sum_{k=1}^{n} a_{ik}b_{ki}, \qquad e_{kk} = \sum_{i=1}^{n} b_{ki}a_{ik}$$

Hence it follows that

$$\text{tr}(AB) = \sum_{i=1}^{n} c_{ii} = \sum_{i=1}^{n}\sum_{k=1}^{n} a_{ik}b_{ki} = \sum_{k=1}^{n}\sum_{i=1}^{n} b_{ki}a_{ik} = \sum_{k=1}^{n} e_{kk} = \text{tr}(BA)$$

(b) If B is similar to A, then $B = S^{-1}AS$. It follows from part (a) that

$$\text{tr}(B) = \text{tr}(S^{-1}(AS)) = \text{tr}((AS)S^{-1}) = \text{tr}(A)$$

MATLAB EXERCISES

2. (a) To determine the matrix representation of L with respect to E set

$$B = U^{-1}AU$$

(b) To determine the matrix representation of L with respect to F set

$$C = V^{-1}AV$$

(c) If B and C are both similar to A then they must be similar to each other. Indeed the transition matrix S from F to E is given by $S = U^{-1}V$ and

$$C = S^{-1}BS$$

CHAPTER TEST A

1. The statement is false in general. If $L: R^n \to R^m$ has matrix representation A and the rank of A is less than n, then it is possible to find vectors $\mathbf{x}_1$ and $\mathbf{x}_2$ such that $L(\mathbf{x}_1) = L(\mathbf{x}_2)$ and $\mathbf{x}_1 \neq \mathbf{x}_2$. For example if
$$A = \begin{pmatrix} 1 & 1 \\ 2 & 2 \end{pmatrix}, \quad \mathbf{x}_1 = \begin{pmatrix} 1 \\ 4 \end{pmatrix}, \quad \mathbf{x}_2 = \begin{pmatrix} 2 \\ 3 \end{pmatrix}$$
and $L: R^2 \to R^2$ is defined by $L(\mathbf{x}) = A\mathbf{x}$, then
$$L(\mathbf{x}_1) = A\mathbf{x}_1 = \begin{pmatrix} 5 \\ 10 \end{pmatrix} = A\mathbf{x}_2 = L(\mathbf{x}_2)$$

2. The statement is true. If $\mathbf{v}$ is any vector in V and c is any scalar, then
$$\begin{aligned}(L_1 + L_2)(c\mathbf{v}) &= L_1(c\mathbf{v}) + L_2(c\mathbf{v}) \\ &= cL_1(\mathbf{v}) + cL_2(\mathbf{v}) \\ &= c(L_1(\mathbf{v}) + L_2(\mathbf{v})) \\ &= c(L_1 + L_2)(\mathbf{v})\end{aligned}$$
If $\mathbf{v}_1$ and $\mathbf{v}_2$ are any vectors in V, then
$$\begin{aligned}(L_1 + L_2)(\mathbf{v}_1 + \mathbf{v}_2) &= L_1(\mathbf{v}_1 + \mathbf{v}_2) + L_2(\mathbf{v}_1 + \mathbf{v}_2) \\ &= L_1(\mathbf{v}_1) + L_1(\mathbf{v}_2) + L_2(\mathbf{v}_1) + L_2(\mathbf{v}_2) \\ &= (L_1(\mathbf{v}_1) + L_2(\mathbf{v}_1)) + (L_1(\mathbf{v}_2) + L_2(\mathbf{v}_2)) \\ &= (L_1 + L_2)(\mathbf{v}_1) + (L_1 + L_2)(\mathbf{v}_2)\end{aligned}$$

3. The statement is true. If $\mathbf{x}$ is in the kernel of L, then $L(\mathbf{x}) = \mathbf{0}$. Thus if $\mathbf{v}$ is any vector in V, then
$$L(\mathbf{v} + \mathbf{x}) = L(\mathbf{v}) + L(\mathbf{x}) = L(\mathbf{v}) + \mathbf{0} = L(\mathbf{v})$$

4. The statement is false in general. To see that $L_1 \neq L_2$, look at the effect of both operators on $\mathbf{e}_1$.
$$L_1(\mathbf{e}_1) = \begin{pmatrix} \frac{1}{2} \\ -\frac{\sqrt{3}}{2} \end{pmatrix} \quad \text{and} \quad L_2(\mathbf{e}_1) = \begin{pmatrix} \frac{1}{2} \\ \frac{\sqrt{3}}{2} \end{pmatrix}$$

5. The statement is false. The set of vectors in the homogeneous coordinate system does not form a subspace of R^3 since it is not closed under addition. If $\mathbf{x}_1$ and $\mathbf{x}_2$ are vectors in the homogeneous system and $\mathbf{y} = \mathbf{x}_1 + \mathbf{x}_2$, then $\mathbf{y}$ is not a vector in the homogeneous coordinate system since $y_3 = 2$.

6. The statement is true. If A is the standard matrix representation of L, then
$$L^2(\mathbf{x}) = L(L(\mathbf{x})) = L(A\mathbf{x}) = A(A\mathbf{x}) = A^2\mathbf{x}$$
for any $\mathbf{x}$ in R^2. Clearly L^2 is a linear transformation since it can be represented by the matrix A^2.

7. The statement is true. If $\mathbf{x}$ is any vector in R^n then it can be represented in terms of the vectors of E
$$\mathbf{x} = c_1\mathbf{x}_{+}c_2\mathbf{x}_2 + \cdots + c_n\mathbf{x}_n$$

If L_1 and L_2 are both represented by the same matrix A with respect to E, then

$$L_1(\mathbf{x}) = d_1\mathbf{x}_+ d_2\mathbf{x}_2 + \cdots + d_n\mathbf{x}_n = L_2(\mathbf{x})$$

where $\mathbf{d} = A\mathbf{c}$. Since $L_1(\mathbf{x}) = L_2(\mathbf{x})$ for all $\mathbf{x} \in R^n$, it follows that $L_1 = L_2$.

8. The statement is true. See Theorem 4.3.1.

9. The statement is true. If A is similar to B and B is similar to C, then there exist nonsingular matrices X and Y such that

$$A = X^{-1}BX \qquad \text{and} \qquad B = Y^{-1}CY$$

If we set $Z = YX$, then Z is nonsingular and

$$A = X^{-1}BX = X^{-1}Y^{-1}CYX = Z^{-1}CZ$$

Thus A is similar to C.

10. The statement is false. Similar matrices have the same trace, but the converse is not true. For example, the matrices

$$B = \begin{pmatrix} 1 & 1 \\ 0 & 1 \end{pmatrix} \qquad \text{and} \qquad I = \begin{pmatrix} 1 & 0 \\ 0 & 1 \end{pmatrix}$$

have trace equal to 2, but the matrices not similar. In fact the only matrix that is similar to the identity matrix is I itself. (If S any nonsingular matrix, then $S^{-1}IS = I$.)

CHAPTER TEST B

1. (a) L is a linear operator since

$$L(c\mathbf{x}) = \begin{pmatrix} cx_1 + cx_2 \\ cx_1 \end{pmatrix} = c\begin{pmatrix} x_1 + x_2 \\ x_1 \end{pmatrix} = cL(\mathbf{x})$$

and

$$L(\mathbf{x}+\mathbf{y}) = \begin{pmatrix} (x_1 + y_1) + (x_2 + y_2) \\ x_1 + y_1 \end{pmatrix} = \begin{pmatrix} x_1 + x_2 \\ x_1 \end{pmatrix} + \begin{pmatrix} y_1 + y_2 \\ y_1 \end{pmatrix}$$
$$= L(\mathbf{x}) + L(\mathbf{y})$$

(b) L is not a letter operator. If, for example we take $\mathbf{x} = (1,1)^T$ then

$$L(2\mathbf{x}) = \begin{pmatrix} 4 \\ 2 \end{pmatrix} \qquad \text{and} \qquad 2L(\mathbf{x}) = \begin{pmatrix} 2 \\ 2 \end{pmatrix}$$

2. To determine the value of $L(\mathbf{v}_3)$ we must first express $\mathbf{v}_3$ as a linear combination of $\mathbf{v}_1$ and $\mathbf{v}_2$. Thus we must find constants c_1 and c_2 such that $\mathbf{v}_3 = c_1\mathbf{v}_1 + c_2\mathbf{v}_2$. In we set $V = (\mathbf{v}_1, \mathbf{v}_2)$ and solve the system $V\mathbf{c} = \mathbf{v}_3$ we see that $\mathbf{c} = (3,2)^T$. It follows then that

$$L(\mathbf{v}_3) = L(3\mathbf{v}_1 + 2\mathbf{v}_2) = 3L(\mathbf{v}_1) + 2L(\mathbf{v}_2) = \begin{pmatrix} 0 \\ 17 \end{pmatrix}$$

3. (a) $\ker(L) = \operatorname{Span}((1,1,1)^T)$
(b) $L(S) = \operatorname{Span}((-1,1,0)^T)$

4. Since

$$L(\mathbf{x}) = \begin{pmatrix} x_2 \\ x_1 \\ x_1 + x_2 \end{pmatrix} = x_1 \begin{pmatrix} 0 \\ 1 \\ 1 \end{pmatrix} + x_2 \begin{pmatrix} 1 \\ 0 \\ 1 \end{pmatrix}$$

it follows that the range of L is the span of the vectors

$$\mathbf{y}_1 = \begin{pmatrix} 0 \\ 1 \\ 1 \end{pmatrix}, \quad \mathbf{y}_2 = \begin{pmatrix} 1 \\ 0 \\ 1 \end{pmatrix}$$

5. Let $\mathbf{e}_1$ and $\mathbf{e}_2$ be the standard basis vectors for R^2. To determine the matrix representation of L we set

$$\mathbf{a}_1 = L(\mathbf{e}_1) = \begin{pmatrix} 1 \\ 1 \\ 3 \end{pmatrix}, \quad \mathbf{a}_2 = L(\mathbf{e}_2) = \begin{pmatrix} 1 \\ -1 \\ 2 \end{pmatrix}$$

If we set

$$A = \begin{pmatrix} 1 & 1 \\ 1 & -1 \\ 3 & 2 \end{pmatrix}$$

then $L(\mathbf{x}) = A\mathbf{x}$ for all $\mathbf{x} \in R^2$.

6. To determine the matrix representation we set

$$\mathbf{a}_1 = L(\mathbf{e}_1) = \begin{pmatrix} -\frac{\sqrt{3}}{2} \\ \frac{1}{2} \end{pmatrix} \quad \text{and} \quad \mathbf{a}_2 = L(\mathbf{e}_2) = \begin{pmatrix} \frac{1}{2} \\ \frac{\sqrt{3}}{2} \end{pmatrix}$$

The matrix representation of the operator is

$$A = (\mathbf{a}_1, \mathbf{a}_2) = \begin{pmatrix} -\frac{\sqrt{3}}{2} & \frac{1}{2} \\ \frac{1}{2} & \frac{\sqrt{3}}{2} \end{pmatrix}$$

7. $A = \begin{pmatrix} 1 & 0 & 5 \\ 0 & 1 & 2 \\ 0 & 0 & 1 \end{pmatrix}$

8. The standard matrix representation for a 45° counterclockwise rotation operator is

$$A = \begin{pmatrix} \cos\frac{\pi}{4} & -\sin\frac{\pi}{4} \\ \sin\frac{\pi}{4} & \cos\frac{\pi}{4} \end{pmatrix} = \begin{pmatrix} \frac{1}{\sqrt{2}} & -\frac{1}{\sqrt{2}} \\ \frac{1}{\sqrt{2}} & \frac{1}{\sqrt{2}} \end{pmatrix}$$

The matrix representation with respect to the basis $[\mathbf{u}_1, \mathbf{u}_2]$ is

$$B = U^{-1}AU = \begin{pmatrix} 2 & -5 \\ -1 & 3 \end{pmatrix} \begin{pmatrix} \frac{1}{\sqrt{2}} & -\frac{1}{\sqrt{2}} \\ \frac{1}{\sqrt{2}} & \frac{1}{\sqrt{2}} \end{pmatrix} \begin{pmatrix} 3 & 5 \\ 1 & 2 \end{pmatrix} = \begin{pmatrix} -\frac{16}{\sqrt{2}} & -\frac{29}{\sqrt{2}} \\ \frac{10}{\sqrt{2}} & \frac{18}{\sqrt{2}} \end{pmatrix}$$

9. **(a)** If $U = (\mathbf{u}_1, \mathbf{u}_2)$ and $V = (\mathbf{v}_1, \mathbf{v}_2)$ then the transition matrix S from $[\mathbf{v}_1, \mathbf{v}_2]$ to $[\mathbf{u}_1, \mathbf{u}_2]$ is

$$S = U^{-1}V = \begin{pmatrix} 2 & -5 \\ -1 & 3 \end{pmatrix} \begin{pmatrix} 1 & 1 \\ -2 & -1 \end{pmatrix} = \begin{pmatrix} 12 & 7 \\ -7 & -4 \end{pmatrix}$$

(b) By Theorem 4.3.1 the matrix representation of L with respect to $[\mathbf{v}_1, \mathbf{v}_2]$ is

$$B = S^{-1}AS = \begin{pmatrix} -4 & -7 \\ 7 & 12 \end{pmatrix} \begin{pmatrix} 2 & 1 \\ 3 & 2 \end{pmatrix} \begin{pmatrix} 12 & 7 \\ -7 & -4 \end{pmatrix} = \begin{pmatrix} -222 & -131 \\ 383 & 226 \end{pmatrix}$$

10. (a) If A and B are similar then $B = S^{-1}AS$ for some nonsingular matrix S. It follows then that

$$\begin{aligned} \det(B) = \det(S^{-1}AS) &= \det(S^{-1})\det(A)\det(S) \\ &= \frac{1}{\det(S)}\det(A)\det(S) = \det(A) \end{aligned}$$

(b) If $B = S^{-1}AS$ then

$$S^{-1}(A - \lambda I)S = S^{-1}AS - \lambda S^{-1}IS = B - \lambda I$$

Therefore $A - \lambda I$ and $B - \lambda I$ ate similar and it follows from part (a) that their determinants must be equal.

Chapter 5

SECTION 1

1. (c) $\cos\theta = \dfrac{14}{\sqrt{221}}, \qquad \theta \approx 10.65°$

(d) $\cos\theta = \dfrac{4\sqrt{6}}{21}, \qquad \theta \approx 62.19°$

3. (b) $\mathbf{p} = (4,4)^T, \ \ \mathbf{x} - \mathbf{p} = (-1,1)^T$

$\mathbf{p}^T(\mathbf{x} - \mathbf{p}) = -4 + 4 = 0$

(d) $\mathbf{p} = (-2,-4,2)^T, \ \ \mathbf{x} - \mathbf{p} = (4,-1,2)^T$

$\mathbf{p}^T(\mathbf{x} - \mathbf{p}) = -8 + 4 + 4 = 0$

4. If $\mathbf{x}$ and $\mathbf{y}$ are linearly independent and θ is the angle between the vectors, then $|\cos\theta| < 1$ and hence

$$|\mathbf{x}^T\mathbf{y}| = \|\mathbf{x}\|\,\|\mathbf{y}\|\,|\cos\theta| < 6$$

8. (b) $-3(x-4) + 6(y-2) + 2(z+5) = 0$

11. (a) $\mathbf{x}^T\mathbf{x} = x_1^2 + x_2^2 \geq 0$

(b) $\mathbf{x}^T\mathbf{y} = x_1y_1 + x_2y_2 = y_1x_1 + y_2x_2 = \mathbf{y}^T\mathbf{x}$

(c) $\mathbf{x}^T(\mathbf{y}+\mathbf{z}) = x_1(y_1+z_1)+x_2(y_2+z_2)$
$= (x_1y_1+x_2y_2)+(x_1z_2+x_2z_2)$
$= \mathbf{x}^T\mathbf{y}+\mathbf{x}^T\mathbf{z}$

12. The inequality can be proved using the Cauchy-Schwarz inequality as follows:

$$\begin{aligned}\|\mathbf{u}+\mathbf{v}\|^2 &= (\mathbf{u}+\mathbf{v})^T(\mathbf{u}+\mathbf{v})\\ &= \mathbf{u}^T\mathbf{u}+\mathbf{v}^T\mathbf{u}+\mathbf{u}^T\mathbf{v}+\mathbf{v}^T\mathbf{v}\\ &= \|\mathbf{u}\|^2+2\mathbf{u}^T\mathbf{v}+\|\mathbf{v}\|^2\\ &= \|\mathbf{u}\|^2+2\|\mathbf{u}\|\,\|\mathbf{v}\|\,\cos\theta+\|\mathbf{v}\|^2\\ &\le \|\mathbf{u}\|^2+2\|\mathbf{u}\|\,\|\mathbf{v}\|+\|\mathbf{v}\|^2\\ &= (\|\mathbf{u}\|+\|\mathbf{v}\|)^2\end{aligned}$$

Taking square roots, we get

$$\|\mathbf{u}+\mathbf{v}\| \le \|\mathbf{u}\|+\|\mathbf{v}\|$$

Equality will hold if and only if $\cos\theta = 1$. This will happen if one of the vectors is a multiple of the other. Geometrically one can think of $\|\mathbf{u}\|$ and $\|\mathbf{v}\|$ as representing the lengths of two sides of a triangle. The length of the third side of the triangle will be $\|\mathbf{u}+\mathbf{v}\|$. Clearly the length of the third side must be less than the sum of the lengths of the first two sides. In the case of equality the triangle degenerates to a line segment.

13. No. For example, if $\mathbf{x}_1 = \mathbf{e}_1$, $\mathbf{x}_2 = \mathbf{e}_2$, $\mathbf{x}_3 = 2\mathbf{e}_1$, then $\mathbf{x}_1 \perp \mathbf{x}_2$, $\mathbf{x}_2 \perp \mathbf{x}_3$, but $\mathbf{x}_1$ is not orthogonal to $\mathbf{x}_3$.

14. (a) By the Pythagorean Theorem

$$\alpha^2+h^2 = \|\mathbf{a}_1\|^2$$

where α is the scalar projection of $\mathbf{a}_1$ onto $\mathbf{a}_2$. It follows that

$$\alpha^2 = \frac{(\mathbf{a}_1^T\mathbf{a}_2)^2}{\|\mathbf{a}_2\|^2}$$

and

$$h^2 = \|\mathbf{a}_1\|^2 - \frac{(\mathbf{a}_1^T\mathbf{a}_2)^2}{\|\mathbf{a}_2\|^2}$$

Hence

$$h^2\|\mathbf{a}_2\|^2 = \|\mathbf{a}_1\|^2\,\|\mathbf{a}_2\|^2 - (\mathbf{a}_1^T\mathbf{a}_2)^2$$

(b) If $\mathbf{a}_1 = (a_{11}, a_{21})^T$ and $\mathbf{a}_2 = (a_{12}, a_{22})^T$, then by part (a)

$$\begin{aligned}h^2\|\mathbf{a}_2\|^2 &= (a_{11}^2+a_{21}^2)(a_{12}^2+a_{22}^2) - (a_{11}a_{12}+a_{21}a_{22})^2\\ &= (a_{11}^2a_{22}^2 - 2a_{11}a_{22}a_{12}a_{21} + a_{21}^2a_{12}^2)\\ &= (a_{11}a_{22}-a_{21}a_{12})^2\end{aligned}$$

Therefore

$$\text{Area of } P = h\|\mathbf{a}_2\| = |a_{11}a_{22}-a_{21}a_{12}| = |\det(A)|$$

15. (a) It θ is the angle between $\mathbf{x}$ and $\mathbf{y}$, then

$$\cos\theta = \frac{\mathbf{x}^T\mathbf{y}}{\|\mathbf{x}\|\,\|\mathbf{y}\|} = \frac{20}{8\cdot 5} = \frac{1}{2}, \qquad \theta = \frac{\pi}{3}$$

(b) The distance between the vectors is given by

$$\|\mathbf{x}-\mathbf{y}\| = \sqrt{0^2+2^2+(-6)^2+3^2} = 7$$

16. (a) Let

$$\alpha = \frac{\mathbf{x}^T\mathbf{y}}{\mathbf{y}^T\mathbf{y}} \qquad \text{and} \qquad \beta = \frac{(\mathbf{x}^T\mathbf{y})^2}{\mathbf{y}^T\mathbf{y}}$$

In terms of these scalars we have $\mathbf{p} = \alpha\mathbf{y}$ and $\mathbf{p}^T\mathbf{x} = \beta$. Furthermore

$$\mathbf{p}^T\mathbf{p} = \alpha^2\mathbf{y}^T\mathbf{y} = \beta$$

and hence

$$\mathbf{p}^T\mathbf{z} = \mathbf{p}^T\mathbf{x} - \mathbf{p}^T\mathbf{p} = \beta - \beta = 0$$

(b) If $\|\mathbf{p}\| = 6$ and $\|\mathbf{z}\| = 8$, then we can apply the Pythagorean law to determine the length of $\mathbf{x} = \mathbf{p} + \mathbf{z}$. It follows that

$$\|\mathbf{x}\|^2 = \|\mathbf{p}\|^2 + \|\mathbf{z}\|^2 = 36 + 64 = 100$$

and hence $\|\mathbf{x}\| = 10$.

17. The matrix Q is unchanged and the nonzero entries of our new search vector $\mathbf{x}$ are $x_6 = \frac{\sqrt{6}}{3}$, $x_7 = \frac{\sqrt{6}}{6}$, $x_{10} = \frac{\sqrt{6}}{6}$. Rounded to three decimal places the search vector is

$$\mathbf{x} = (0,\ 0,\ 0,\ 0,\ 0,\ 0.816,\ 0.408,\ 0,\ 0,\ 0.408)^T$$

The search results are given by the vector

$$\mathbf{y} = Q^T\mathbf{x} = (0,\ 0.161,\ 0.401\ 0.234,\ 0.612,\ 0.694,\ 0,\ 0.504)^T$$

The largest entry of $\mathbf{y}$ is $y_6 = 0.694$. This implies that Module 6 is the one that best meets our search criteria.

SECTION 2

1. (b) The reduced row echelon form of A is

$$\begin{pmatrix} 1 & 0 & -2 \\ 0 & 1 & 1 \end{pmatrix}$$

The set $\{(2,\ -1,\ 1)^T\}$ is a basis for $N(A)$ and $\{(1,\ 0,\ -2)^T,\ (0,\ 1,\ 1)^T\}$ is a basis for $R(A^T)$. The reduced row echelon form of A^T is

$$\begin{pmatrix} 1 & 0 \\ 0 & 1 \\ 0 & 0 \end{pmatrix}$$

$N(A^T) = \{(0,\ 0)^T\}$ and $\{(1,\ 0)^T,\ (0,\ 1)^T\}$ is a basis for $R(A) = R^2$.

(c) The reduced row echelon form of A is

$$\begin{pmatrix} 1 & 0 \\ 0 & 1 \\ 0 & 0 \\ 0 & 0 \end{pmatrix}$$

$N(A) = \{(0,\ 0)^T\}$ and $\{(1,\ 0)^T,\ (0,\ 1)^T\}$ is a basis for $R(A^T)$. The reduced row echelon form of A^T is

$$U = \begin{pmatrix} 1 & 0 & \frac{5}{14} & \frac{5}{14} \\ 0 & 1 & \frac{4}{7} & \frac{11}{7} \end{pmatrix}$$

We can obtain a basis for $R(A)$ by transposing the rows of U and we can obtain a basis for $N(A^T)$ by solving $U\mathbf{x} = \mathbf{0}$. It follows that

$$\left\{ \begin{pmatrix} 1 \\ 0 \\ \frac{5}{14} \\ \frac{5}{14} \end{pmatrix}, \begin{pmatrix} 0 \\ 1 \\ \frac{4}{7} \\ \frac{11}{7} \end{pmatrix} \right\} \quad \text{and} \quad \left\{ \begin{pmatrix} -\frac{5}{14} \\ -\frac{4}{7} \\ 1 \\ 0 \end{pmatrix}, \begin{pmatrix} -\frac{5}{14} \\ -\frac{11}{7} \\ 0 \\ 1 \end{pmatrix} \right\}$$

are bases for $R(A)$ and $N(A^T)$, respectively.

2. (b) S corresponds to a line ℓ in 3-space that passes through the origin and the point $(1,\ -1,\ 1)$. $S^\perp$ corresponds to a plane in 3-space that passes through the origin and is normal to the line ℓ.

3. (a) A vector $\mathbf{z}$ will be in $S^\perp$ if and only if $\mathbf{z}$ is orthogonal to both $\mathbf{x}$ and $\mathbf{y}$. Since $\mathbf{x}^T$ and $\mathbf{y}^T$ are the row vectors of A, it follows that $S^\perp = N(A)$.

6. No. $(3,\ 1,\ 2)^T$ and $(2,\ 1,\ 1)^T$ are not orthogonal.

7. No. Since $N(A^T)$ and $R(A)$ are orthogonal complements

$$N(A^T) \cap R(A) = \{\mathbf{0}\}$$

The vector $\mathbf{a}_j$ cannot be in $N(A^T)$ since it is a nonzero element of $R(A)$. Also, note that the jth coordinate of $A^T\mathbf{a}_j$ is

$$\mathbf{a}_j^T\mathbf{a}_j = \|\mathbf{a}_j\|^2 > 0$$

8. If $\mathbf{y} \in S^\perp$ then since each $\mathbf{x}_i \in S$ it follows that $\mathbf{y} \perp \mathbf{x}_i$ for $i = 1, \ldots, k$. Conversely if $\mathbf{y} \perp \mathbf{x}_i$ for $i = 1, \ldots, k$ and $\mathbf{x} = \alpha_1\mathbf{x}_1 + \alpha_2\mathbf{x}_2 + \cdots + \alpha_k\mathbf{x}_k$ is any element of S, then

$$\mathbf{y}^T\mathbf{x} = \mathbf{y}^T \left(\sum_{i=1}^{k} \alpha_i \mathbf{x}_i \right) = \sum_{i=1}^{k} \alpha_i \mathbf{y}^T\mathbf{x}_i = 0$$

Thus $\mathbf{y} \in S^\perp$.

10. **Corollary 5.2.5.** If A is an $m \times n$ matrix and $\mathbf{b} \in R^n$, then either there is a vector $\mathbf{x} \in R^n$ such that $A\mathbf{x} = \mathbf{b}$ or there is a vector $\mathbf{y} \in R^m$ such that $A^T\mathbf{y} = \mathbf{0}$ and $\mathbf{y}^T\mathbf{b} \neq 0$.

Proof: If $A\mathbf{x} = \mathbf{b}$ has no solution then $\mathbf{b} \notin R(A)$. Since $R(A) = N(A^T)^\perp$ it follows that $\mathbf{b} \notin N(A^T)^\perp$. But this means that there is a vector $\mathbf{y}$ in $N(A^T)$ that is not orthogonal to $\mathbf{b}$. Thus $A^T\mathbf{y} = \mathbf{0}$ and $\mathbf{y}^T\mathbf{b} \neq 0$.

11. If $\mathbf{x}$ is not a solution to $A\mathbf{x} = \mathbf{0}$ then $\mathbf{x} \notin N(A)$. Since $N(A) = R(A^T)^\perp$ it follows that $\mathbf{x} \notin R(A^T)^\perp$. Thus there exists a vector $\mathbf{y}$ in $R(A^T)$ that is not orthogonal to $\mathbf{x}$, i.e., $\mathbf{x}^T\mathbf{y} \neq \mathbf{0}$.

12. Part (a) follows since $R^n = N(A) \oplus R(A^T)$.
Part (b) follows since $R^m = N(A^T) \oplus R(A)$.

13. (a) $A\mathbf{x} \in R(A)$ for all vectors $\mathbf{x}$ in R^n. If $\mathbf{x} \in N(A^TA)$ then

$$A^TA\mathbf{x} = \mathbf{0}$$

and hence $A\mathbf{x} \in N(A^T)$.

(b) If $\mathbf{x} \in N(A)$, then

$$A^TA\mathbf{x} = A^T\mathbf{0} = \mathbf{0}$$

and hence $\mathbf{x} \in N(A^TA)$. Thus $N(A)$ is a subspace of $N(A^TA)$.

Conversely, if $\mathbf{x} \in N(A^TA)$, then by part (a), $A\mathbf{x} \in R(A) \cap N(A^T)$. Since $R(A) \cap N(A^T) = \{\mathbf{0}\}$, it follows that $\mathbf{x} \in N(A)$. Thus $N(A^TA)$ is a subspace of $N(A)$. It follows then that $N(A^TA) = N(A)$.

(c) A and A^TA have the same nullspace and consequently must have the same nullity. Since both matrices have n columns, it follows from the Rank-Nullity Theorem that they must also have the same rank.

(d) If A has linearly independent columns then A has rank n. By part (c), A^TA also has rank n and consequently is nonsingular.

14. (a) If $\mathbf{x} \in N(B)$, then

$$C\mathbf{x} = AB\mathbf{x} = A\mathbf{0} = \mathbf{0}$$

Thus $\mathbf{x} \in N(C)$ and it follows that $N(B)$ is a subspace of $N(C)$.

(b) If $\mathbf{x} \in N(C)^\perp$, then $\mathbf{x}^T\mathbf{y} = 0$ for all $\mathbf{y} \in N(C)$. Since $N(B) \subset N(C)$ it follows that $\mathbf{x}$ is orthogonal to each element of $N(B)$ and hence $\mathbf{x} \in N(B)^\perp$. Therefore

$$R(C^T) = N(C)^\perp \text{ is a subspace of } N(B)^\perp = R(B^T)$$

15. Let $\mathbf{x} \in U \cap V$. We can write

$$\begin{array}{lll} \mathbf{x} = \mathbf{0} + \mathbf{x} & (\mathbf{0} \in U, & \mathbf{x} \in V) \\ \mathbf{x} = \mathbf{x} + \mathbf{0} & (\mathbf{x} \in U, & \mathbf{0} \in V) \end{array}$$

By the uniqueness of the direct sum representation $\mathbf{x} = \mathbf{0}$.

16. It was shown in the text that

$$R(A) = \{A\mathbf{y} \mid \mathbf{y} \in R(A^T)\}$$

If $\mathbf{y} \in R(A^T)$, then we can write

$$\mathbf{y} = \alpha_1\mathbf{x}_1 + \alpha_2\mathbf{x}_2 + \cdots + \alpha_r\mathbf{x}_r$$

Thus

$$A\mathbf{y} = \alpha_1 A\mathbf{x}_1 + \alpha_2 A\mathbf{x}_2 + \cdots + \alpha_r A\mathbf{x}_r$$

and it follows that the vectors $A\mathbf{x}_1, \ldots, A\mathbf{x}_r$ span $R(A)$. Since $R(A)$ has dimension r, $\{A\mathbf{x}_1, \ldots, A\mathbf{x}_r\}$ is a basis for $R(A)$.

17. **(a)** A is symmetric since

$$\begin{aligned} A^T &= (\mathbf{x}\mathbf{y}^T + \mathbf{y}\mathbf{x}^T)^T = (\mathbf{x}\mathbf{y}^T)^T + (\mathbf{y}\mathbf{x}^T)^T \\ &= (\mathbf{y}^T)^T\mathbf{x}^T + (\mathbf{x}^T)^T\mathbf{y}^T = \mathbf{y}\mathbf{x}^T + \mathbf{x}\mathbf{y}^T = A \end{aligned}$$

(b) For any vector $\mathbf{z}$ in R^n

$$A\mathbf{z} = \mathbf{x}\mathbf{y}^T\mathbf{z} + \mathbf{y}\mathbf{x}^T\mathbf{z} = c_1\mathbf{x} + c_2\mathbf{y}$$

where $c_1 = \mathbf{y}^T\mathbf{z}$ and $c_2 = \mathbf{x}^T\mathbf{z}$. If $\mathbf{z}$ is in $N(A)$ then

$$\mathbf{0} = A\mathbf{z} = c_1\mathbf{x} + c_2\mathbf{y}$$

and since $\mathbf{x}$ and $\mathbf{y}$ are linearly independent we have $\mathbf{y}^T\mathbf{z} = c_1 = 0$ and $\mathbf{x}^T\mathbf{z} = c_2 = 0$. So $\mathbf{z}$ is orthogonal to both $\mathbf{x}$ and $\mathbf{y}$. Since $\mathbf{x}$ and $\mathbf{y}$ span S it follows that $\mathbf{z} \in S^\perp$.

Conversely, if $\mathbf{z}$ is in $S^\perp$ then $\mathbf{z}$ is orthogonal to both $\mathbf{x}$ and $\mathbf{y}$. It follows that

$$A\mathbf{z} = c_1\mathbf{x} + c_2\mathbf{y} = \mathbf{0}$$

since $c_1 = \mathbf{y}^T\mathbf{z} = 0$ and $c_2 = \mathbf{x}^T\mathbf{z} = 0$. Therefore $\mathbf{z}$ is in $N(A)$ and hence $N(A) = S^\perp$.

(c) Clearly $\dim S = 2$ and by Theorem 5.2.2, $\dim S + \dim S^\perp = n$. Using our result from part (a) we have

$$\dim N(A) = \dim S^\perp = n - 2$$

So A has nullity $n - 2$. It follows from the Rank-Nullity Theorem that the rank of A must be 2.

SECTION 3

1. (b) $A^TA = \begin{pmatrix} 6 & -1 \\ -1 & 6 \end{pmatrix}$ and $A^T\mathbf{b} = \begin{pmatrix} 20 \\ -25 \end{pmatrix}$

The solution to the normal equations $A^TA\mathbf{x} = A^T\mathbf{b}$ is

$$\mathbf{x} = \begin{pmatrix} 19/7 \\ -26/7 \end{pmatrix}$$

2. (Exercise 1b.)

(a) $\mathbf{p} = \frac{1}{7}(-45,\ 12,\ 71)^T$

(b) $\mathbf{r} = \frac{1}{7}(115,\ 23,\ 69)^T$

(c)

$$A^T\mathbf{r} = \begin{pmatrix} -1 & 2 & 1 \\ 1 & 1 & -2 \end{pmatrix} \begin{pmatrix} \frac{115}{7} \\ \frac{23}{7} \\ \frac{69}{7} \end{pmatrix} = \begin{pmatrix} 0 \\ 0 \\ 0 \end{pmatrix}$$

Therefore $\mathbf{r}$ is in $N(A^T)$.

6. $A = \begin{pmatrix} 1 & -1 & 1 \\ 1 & 0 & 0 \\ 1 & 1 & 1 \\ 1 & 2 & 4 \end{pmatrix}, \quad \mathbf{b} = \begin{pmatrix} 0 \\ 1 \\ 3 \\ 9 \end{pmatrix}$

$$A^TA = \begin{pmatrix} 4 & 2 & 6 \\ 2 & 6 & 8 \\ 6 & 8 & 18 \end{pmatrix}, \quad A^T\mathbf{b} = \begin{pmatrix} 13 \\ 21 \\ 39 \end{pmatrix}$$

The solution to $A^TA\mathbf{x} = A^T\mathbf{b}$ is $(0.6,\ 1.7,\ 1.2)^T$. Therefore the best least squares fit by a quadratic polynomial is given by

$$p(x) = 0.6 + 1.7x + 1.2x^2$$

7. To find the best fit by a linear function we must find the least squares solution to the linear system

$$\begin{pmatrix} 1 & x_1 \\ 1 & x_2 \\ \vdots & \vdots \\ 1 & x_n \end{pmatrix} \begin{pmatrix} c_0 \\ c_1 \end{pmatrix} = \begin{pmatrix} y_1 \\ y_2 \\ \vdots \\ y_n \end{pmatrix}$$

If we form the normal equations the augmented matrix for the system will be

$$\left(\begin{array}{cc|c} n & \sum_{i=1}^{n} x_i & \sum_{i=1}^{n} y_i \\ \sum_{i=1}^{n} x_i & \sum_{i=1}^{n} x_i^2 & \sum_{i=1}^{n} x_i y_i \end{array}\right)$$

If $\overline{x} = 0$ then

$$\sum_{i=1}^{n} x_i = n\overline{x} = 0$$

and hence the coefficient matrix for the system is diagonal. The solution is easily obtained.

$$c_0 = \frac{\sum_{i=1}^{n} y_i}{n} = \overline{y}$$

and

$$c_1 = \frac{\sum_{i=1}^{n} x_i y_i}{\sum_{i=1}^{n} x_i^2} = \frac{\mathbf{x}^T\mathbf{y}}{\mathbf{x}^T\mathbf{x}}$$

8. To show that the least squares line passes through the center of mass, we introduce a new variable $z = x - \overline{x}$. If we set $z_i = x_i - \overline{x}$ for $i = 1, \ldots, n$, then $\overline{z} = 0$. Using the result from Exercise 7 the equation of the best least squares fit by a linear function in the new zy-coordinate system is

$$y = \overline{y} + \frac{\mathbf{z}^T\mathbf{y}}{\mathbf{z}^T\mathbf{z}}z$$

If we translate this back to xy-coordinates we end up with the equation

$$y - \overline{y} = c_1(x - \overline{x})$$

where

$$c_1 = \frac{\sum_{i=1}^{n}(x_i - \overline{x})y_i}{\sum_{i=1}^{n}(x_i - \overline{x})^2}$$

9. (a) If $\mathbf{b} \in R(A)$ then $\mathbf{b} = A\mathbf{x}$ for some $\mathbf{x} \in R^n$. It follows that

$$P\mathbf{b} = PA\mathbf{x} = A(A^TA)^{-1}A^TA\mathbf{x} = A\mathbf{x} = \mathbf{b}$$

(b) If $\mathbf{b} \in R(A)^{\perp}$ then since $R(A)^{\perp} = N(A^T)$ it follows that $A^T\mathbf{b} = \mathbf{0}$ and hence

$$P\mathbf{b} = A(A^TA)^{-1}A^T\mathbf{b} = \mathbf{0}$$

(c) The following figures give a geometric illustration of parts (a) and (b). In the first figure $\mathbf{b}$ lies in the plane corresponding to $R(A)$. Since it is already in the plane, projecting it onto the plane will have no effect. In the second figure $\mathbf{b}$ lies on the line through the origin that is normal to the plane. When it is projected onto the plane it projects right down to the origin.

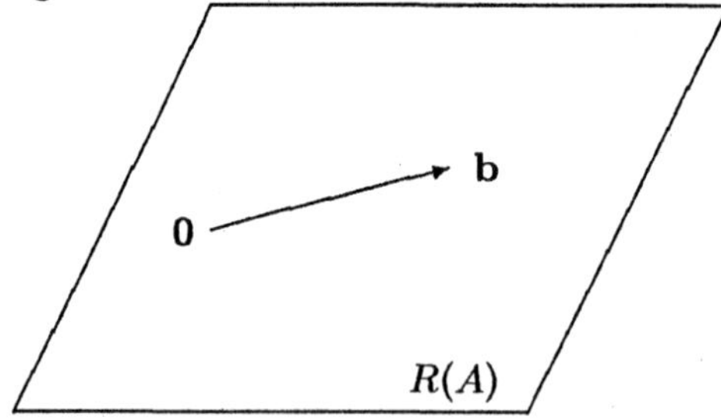

If $\mathbf{b} \in R(A)$, then $P\mathbf{b} = \mathbf{b}$.

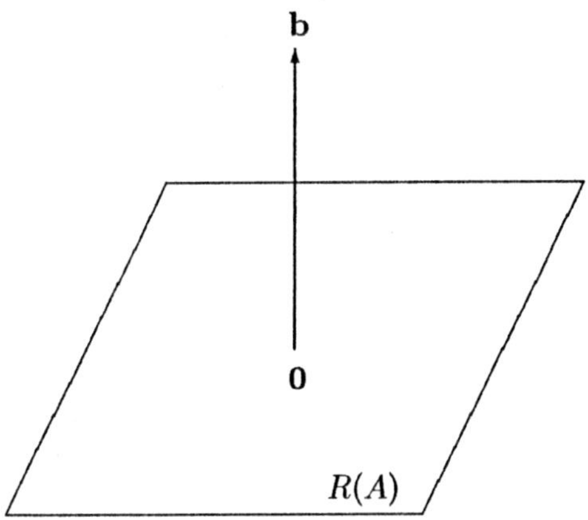

If $\mathbf{b} \in R(A)^{\perp}$, then $P\mathbf{b} = \mathbf{0}$.

10. **(a)** By the Consistency Theorem $A\mathbf{x} = \mathbf{b}$ is consistent if and only if $\mathbf{b}$ is in $R(A)$. We are given that $\mathbf{b}$ is in $N(A^T)$. So if the system is consistent then $\mathbf{b}$ would be in $R(A) \cap N(A^T) = \{\mathbf{0}\}$. Since $\mathbf{b} \neq \mathbf{0}$, the system must be inconsistent.

(b) If A has rank 3 then A^TA also has rank 3 (see Exercise 13 in Section 2). The normal equations are always consistent and in this case there will be 2 free variables. So the least squares problem will have infinitely many solutions.

11. (a) $P^2 = A(A^TA)^{-1}A^TA(A^TA)^{-1}A^T = A(A^TA)^{-1}A^T = P$

(b) Prove: $P^k = P$ for $k = 1, 2, \ldots$.

Proof: The proof is by mathematical induction. In the case $k = 1$ we have $P^1 = P$. If $P^m = P$ for some m then

$$P^{m+1} = PP^m = PP = P^2 = P$$

(c)
$$\begin{aligned} P^T &= [A(A^TA)^{-1}A^T]^T \\ &= (A^T)^T[(A^TA)^{-1}]^TA^T \\ &= A[(A^TA)^T]^{-1}A^T \\ &= A(A^TA)^{-1}A^T \\ &= P \end{aligned}$$

12. If

$$\begin{pmatrix} A & I \\ O & A^T \end{pmatrix} \begin{pmatrix} \hat{\mathbf{x}} \\ \mathbf{r} \end{pmatrix} = \begin{pmatrix} \mathbf{b} \\ \mathbf{0} \end{pmatrix}$$

then

$$\begin{aligned} A\hat{\mathbf{x}} + \mathbf{r} &= \mathbf{b} \\ A^T\mathbf{r} &= \mathbf{0} \end{aligned}$$

We have then that

$$\mathbf{r} = \mathbf{b} - A\hat{\mathbf{x}}$$

$$A^T\mathbf{r} = A^T\mathbf{b} - A^TA\hat{\mathbf{x}} = \mathbf{0}$$

Therefore

$$A^TA\hat{\mathbf{x}} = A^T\mathbf{b}$$

So $\hat{\mathbf{x}}$ is a solution to the normal equations and hence is the least squares solution to $A\mathbf{x} = \mathbf{b}$.

13. If $\hat{\mathbf{x}}$ is a solution to the least squares problem, then $\hat{\mathbf{x}}$ is a solution to the normal equations

$$A^TA\mathbf{x} = A^T\mathbf{b}$$

It follows that a vector $\mathbf{y} \in R^n$ will be a solution if and only if

$$\mathbf{y} = \hat{\mathbf{x}} + \mathbf{z}$$

for some $\mathbf{z} \in N(A^TA)$. (See Exercise 20, Chapter 3, Section 6). Since

$$N(A^TA) = N(A)$$

we conclude that $\mathbf{y}$ is a least squares solution if and only if

$$\mathbf{y} = \hat{\mathbf{x}} + \mathbf{z}$$

for some $\mathbf{z} \in N(A)$.

SECTION 4

2. (b) $\mathbf{p} = \dfrac{\mathbf{x}^T\mathbf{y}}{\mathbf{y}^T\mathbf{y}}\mathbf{y} = \dfrac{12}{72}\mathbf{y} = \left(\dfrac{4}{3}, \dfrac{1}{3}, \dfrac{1}{3}, 0\right)^T$

(c) $\mathbf{x} - \mathbf{p} = \left(-\dfrac{1}{3}, \dfrac{2}{3}, \dfrac{2}{3}, 1\right)^T$

$(\mathbf{x} - \mathbf{p})^T\mathbf{p} = -\dfrac{4}{9} + \dfrac{2}{9} + \dfrac{2}{9} + 0 = 0$

(d) $\|\mathbf{x} - \mathbf{p}\|_2 = \sqrt{2}$, $\|\mathbf{p}\|_2 = \sqrt{2}$, $\|\mathbf{x}\|_2 = 2$

$$\|\mathbf{x} - \mathbf{p}\|^2 + \|\mathbf{p}\|^2 = 4 = \|\mathbf{x}\|^2$$

3. (a) $\langle \mathbf{x}, \mathbf{y} \rangle = x_1y_1w_1 + x_2y_2w_2 + x_3y_3w_3 = 1 \cdot -5 \cdot \dfrac{1}{4} + 1 \cdot 1 \cdot \dfrac{1}{2} + 1 \cdot 3 \cdot \dfrac{1}{4} = 0$

5. (i)

$$\langle A, A \rangle = \sum_{i=1}^{m}\sum_{j=1}^{n} a_{ij}^2 \geq 0$$

and $\langle A, A \rangle = 0$ if and only if each $a_{ij} = 0$.

(ii) $\langle A, B \rangle = \displaystyle\sum_{i=1}^{m}\sum_{j=1}^{n} a_{ij}b_{ij} = \sum_{i=1}^{m}\sum_{j=1}^{n} b_{ij}a_{ij} = \langle B, A \rangle$

(iii)

$$\begin{aligned}\langle \alpha A + \beta B, C\rangle &= \sum_{i=1}^{m}\sum_{j=1}^{n}(\alpha a_{ij} + \beta b_{ij})c_{ij}\\ &= \alpha\sum_{i=1}^{m}\sum_{j=1}^{n} a_{ij}c_{ij} + \beta\sum_{i=1}^{m}\sum_{j=1}^{n} b_{ij}c_{ij}\\ &= \alpha\langle A, C\rangle + \beta\langle B, C\rangle\end{aligned}$$

6. Show that the inner product on $C[a, b]$ determined by

$$\langle f, g\rangle = \int_a^b f(x)g(x)\,dx$$

satisfies the last two conditions of the definition of an inner product.

Solution:

(ii) $\langle f, g\rangle = \int_a^b f(x)g(x)\,dx = \int_a^b g(x)f(x)\,dx = \langle g, f\rangle$

(iii)
$$\begin{aligned}\langle \alpha f + \beta g, h\rangle &= \int_a^b (\alpha f(x) + \beta g(x))h(x)\,dx\\ &= \alpha\int_a^b f(x)h(x)\,dx + \beta\int_a^b g(x)h(x)\,dx\\ &= \alpha\langle f, h\rangle + \beta\langle g, h\rangle\end{aligned}$$

7 (c)

$$\langle x^2, x^3\rangle = \int_0^1 x^2x^3dx = \frac{1}{6}$$

8 (c)

$$\|1\|^2 = \int_0^1 1\cdot 1\,dx = 1$$

$$\|\mathbf{p}\|^2 = \int_0^1 \frac{9}{4}x^2\,dx = \frac{3}{4}$$

$$\|1 - \mathbf{p}\|^2 = \int_0^1 \left(1 - \frac{3}{2}x\right)^2 dx = \frac{1}{4}$$

Thus $\|1\| = 1$, $\|\mathbf{p}\| = \frac{\sqrt{3}}{2}$, $\|1 - \mathbf{p}\| = \frac{1}{2}$, and

$$\|1 - \mathbf{p}\|^2 + \|\mathbf{p}\|^2 = 1 = \|1\|^2$$

9. The vectors $\cos mx$ and $\sin nx$ are orthogonal since

$$\begin{aligned}\langle \cos mx, \sin nx\rangle &= \frac{1}{\pi}\int_{-\pi}^{\pi} \cos mx \sin nx\,dx\\ &= \frac{1}{2\pi}\int_{-\pi}^{\pi} [\sin(n+m)x + \sin(n-m)x]\,dx\\ &= 0\end{aligned}$$

They are unit vectors since

$$\begin{aligned}\langle \cos mx, \cos mx\rangle &= \frac{1}{\pi}\int_{-\pi}^{\pi} \cos^2 mx\,dx \\ &= \frac{1}{2\pi}\int_{-\pi}^{\pi} [1+\cos 2mx]\,dx \\ &= 1\end{aligned}$$

$$\begin{aligned}\langle \sin nx, \sin nx\rangle &= \frac{1}{\pi}\int_{-\pi}^{\pi} \sin nx \sin nx\,dx \\ &= \frac{1}{2\pi}\int_{-\pi}^{\pi} (1-\cos 2nx)\,dx \\ &= 1\end{aligned}$$

Since the $\cos mx$ and $\sin nx$ are orthogonal, the distance between the vectors can be determined using the Pythagorean law.

$$\|\cos mx - \sin nx\| = (\|\cos mx\|^2 + \|\sin nx\|^2)^{\frac{1}{2}} = \sqrt{2}$$

10. $\langle x, x^2\rangle = \sum_{i=1}^{5} x_i x_i^2 = -1 - \frac{1}{8} + 0 + \frac{1}{8} + 1 = 0$

11. (c) $\|x - x^2\| = \left(\sum_{i=1}^{5}(x_i - x_i^2)^2\right)^{1/2} = \frac{\sqrt{26}}{4}$

12. (i) By the definition of an inner product we have $\langle \mathbf{v}, \mathbf{v}\rangle \geq 0$ with equality if and only if $\mathbf{v} = \mathbf{0}$. Thus $\|\mathbf{v}\| = \sqrt{\langle \mathbf{v}, \mathbf{v}\rangle} \geq 0$ and $\|\mathbf{v}\| = 0$ if and only if $\mathbf{v} = \mathbf{0}$.

(ii) $\|\alpha\mathbf{v}\| = \sqrt{\langle \alpha\mathbf{v}, \alpha\mathbf{v}\rangle} = \sqrt{\alpha^2\langle \mathbf{v}, \mathbf{v}\rangle} = |\alpha|\,\|\mathbf{v}\|$

13. (i) Clearly

$$\sum_{i=1}^{n} |x_i| \geq 0$$

If

$$\sum_{i=1}^{n} |x_i| = 0$$

then all of the x_i's must be 0.

(ii) $\|\alpha\mathbf{x}\|_1 = \sum_{i=1}^{n} |\alpha x_i| = |\alpha| \sum_{i=1}^{n} |x_i| = |\alpha|\,\|\mathbf{x}\|_1$

(iii) $\|\mathbf{x}+\mathbf{y}\|_1 = \sum_{i=1}^{n} |x_i + y_i| \leq \sum_{i=1}^{n} |x_i| + \sum_{i=1}^{n} |y_i| = \|\mathbf{x}\|_1 + \|\mathbf{y}\|_1$

14. (i) $\|\mathbf{x}\|_\infty = \max_{1\leq i\leq n} |x_i| \geq 0$. If $\max_{1\leq i\leq n} |x_i| = 0$ then all of the x_i's must be zero.

(ii) $\|\alpha\mathbf{x}\|_\infty = \max_{1\leq i\leq n} |\alpha x_i| = |\alpha| \max_{1\leq i\leq n} |x_i| = |\alpha|\,\|\mathbf{x}\|_\infty$

(iii) $\|\mathbf{x}+\mathbf{y}\|_\infty = \max |x_i + y_i| \leq \max |x_i| + \max |y_i| = \|\mathbf{x}\|_\infty + \|\mathbf{y}\|_\infty$

17. If $\langle \mathbf{x}, \mathbf{y} \rangle = 0$, then

$$\begin{aligned} \|\mathbf{x}-\mathbf{y}\|^2 &= \langle \mathbf{x}-\mathbf{y}, \mathbf{x}-\mathbf{y} \rangle \\ &= \langle \mathbf{x}, \mathbf{x} \rangle - 2\langle \mathbf{x}, \mathbf{y} \rangle + \langle \mathbf{y}, \mathbf{y} \rangle \\ &= \|\mathbf{x}\|^2 + \|\mathbf{y}\|^2 \end{aligned}$$

Therefore

$$\|\mathbf{x}-\mathbf{y}\| = (\|\mathbf{x}\|^2 + \|\mathbf{y}\|^2)^{1/2}$$

Alternatively, one can prove this result by noting that if $\mathbf{x}$ is orthogonal to $\mathbf{y}$ then $\mathbf{x}$ is also orthogonal to $-\mathbf{y}$ and hence by the Pythagorean Law

$$\|\mathbf{x}-\mathbf{y}\|^2 = \|\mathbf{x}+(-\mathbf{y})\|^2 = \|\mathbf{x}\|^2 + \|-\mathbf{y}\|^2 = \|\mathbf{x}\|^2 + \|\mathbf{y}\|^2$$

18. $\|\mathbf{x}-\mathbf{y}\| = (\langle \mathbf{x}-\mathbf{y}, \mathbf{x}-\mathbf{y} \rangle)^{1/2} = \left(\sum_{i=1}^{n} (x_i - y_i)^2 \right)^{1/2}$

19. For $i = 1, \ldots, n$

$$|x_i| \le (x_1^2 + x_2^2 + \cdots + x_n^2)^{1/2} = \|\mathbf{x}\|_2$$

Thus

$$\|\mathbf{x}\|_\infty = \max_{1\le i\le n} |x_i| \le \|\mathbf{x}\|_2$$

20. $$\begin{aligned} \|\mathbf{x}\|_2 &= \|x_1\mathbf{e}_1 + x_2\mathbf{e}_2\|_2 \\ &\le \|x_1\mathbf{e}_1\|_2 + \|x_2\mathbf{e}_2\|_2 \\ &= |x_1|\,\|\mathbf{e}_1\|_2 + |x_2|\,\|\mathbf{e}_2\|_2 \\ &= |x_1| + |x_2| \\ &= \|\mathbf{x}\|_1 \end{aligned}$$

21. $\mathbf{e}_1$ and $\mathbf{e}_2$ are both examples.

22. $\|-\mathbf{v}\| = \|(-1)\mathbf{v}\| = |-1|\,\|\mathbf{v}\| = \|\mathbf{v}\|$

23. $$\begin{aligned} \|\mathbf{u}+\mathbf{v}\|^2 &= \langle \mathbf{u}+\mathbf{v}, \mathbf{u}+\mathbf{v} \rangle \\ &= \|\mathbf{u}\|^2 + 2\langle \mathbf{u}, \mathbf{v} \rangle + \|\mathbf{v}\|^2 \\ &\ge \|\mathbf{u}\|^2 - 2\|\mathbf{u}\|\,\|\mathbf{v}\| + \|\mathbf{v}\|^2 \\ &= (\|\mathbf{u}\| - \|\mathbf{v}\|)^2 \end{aligned}$$

24.

$$\begin{aligned} \|\mathbf{u}+\mathbf{v}\|^2 &= \|\mathbf{u}\|^2 + 2\langle \mathbf{u}, \mathbf{v} \rangle + \|\mathbf{v}\|^2 \\ \|\mathbf{u}-\mathbf{v}\|^2 &= \|\mathbf{u}\|^2 - 2\langle \mathbf{u}, \mathbf{v} \rangle + \|\mathbf{v}\|^2 \\ \|\mathbf{u}+\mathbf{v}\|^2 + \|\mathbf{u}-\mathbf{v}\|^2 &= 2\|\mathbf{u}\|^2 + 2\|\mathbf{v}\|^2 = 2(\|\mathbf{u}\|^2 + \|\mathbf{v}\|^2) \end{aligned}$$

If the vectors $\mathbf{u}$ and $\mathbf{v}$ are used to form a parallelogram in the plane, then the diagonals will be $\mathbf{u}+\mathbf{v}$ and $\mathbf{u}-\mathbf{v}$. The equation shows that the sum of the squares of the lengths of the diagonals is twice the sum of the squares of the lengths of the two sides.

25. The result will not be valid for most choices of $\mathbf{u}$ and $\mathbf{v}$. For example, if $\mathbf{u} = \mathbf{e}_1$ and $\mathbf{v} = \mathbf{e}_2$, then

$$\begin{aligned} \|\mathbf{u}+\mathbf{v}\|_1^2 + \|\mathbf{u}-\mathbf{v}\|_1^2 &= 2^2 + 2^2 = 8 \\ 2\|\mathbf{u}\|_1^2 + 2\|\mathbf{v}\|_1^2 &= 2 + 2 = 4 \end{aligned}$$

26. (a) The equation

$$\|f\| = |f(a)| + |f(b)|$$

does not define a norm on $C[a,b]$. For example, the function $f(x) = x^2 - x$ in $C[0,1]$ has the property

$$\|f\| = |f(0)| + |f(1)| = 0$$

however, f is not the zero function.

(b) The expression

$$\|f\| = \int_a^b |f(x)|\, dx$$

defines a norm on $C[a,b]$. To see this we must show that the three conditions in the definition of norm are satisfied.

(i) $\int_a^b |f(x)|\, dx \geq 0$. Equality can occur if and only if f is the zero function. Indeed, if $f(x_0) \neq 0$ for some x_0 in $[a,b]$, then the continuity of $f(x)$ implies that $|f(x)| > 0$ for all x in some interval containing x_0 and consequently $\int_a^b |f(x)|\, dx > 0$.

(ii)

$$\|\alpha f\| = \int_a^b |\alpha f(x)|\, dx = |\alpha| \int_a^b |f(x)|\, dx = |\alpha| \|f\|$$

(iii)

$$\begin{aligned}\|f+g\| &= \int_a^b |f(x)+g(x)|\, dx \\ &\leq \int_a^b (|f(x)| + |g(x)|)\, dx \\ &= \int_a^b |f(x)|\, dx + \int_a^b |g(x)|\, dx \\ &= \|f\| + \|g\|\end{aligned}$$

(c) The expression

$$\|f\| = \max_{a\leq x\leq b} |f(x)|$$

defines a norm on $C[a,b]$. To see this we must verify that three conditions are satisfied.

(i) Clearly $\max_{a\leq x\leq b} |f(x)| \geq 0$. Equality can occur only if f is the zero function.

(ii)

$$\|\alpha f\| = \max_{a\leq x\leq b} |\alpha f(x)| = |\alpha| \max_{a\leq x\leq b} |f(x)| = |\alpha|\, \|f\|$$

(iii)

$$\begin{aligned}\|f+g\| &= \max_{a\leq x\leq b} |f(x)+g(x)| \\ &\leq \max_{a\leq x\leq b} (|f(x)| + |g(x)|)\end{aligned}$$

$$\leq \max_{a\leq x\leq b}|f(x)| + \max_{a\leq x\leq b}|g(x)|$$
$$= \|f\| + \|g\|$$

27. (a) If $\mathbf{x} \in R^n$, then

$$|x_i| \leq \max_{1\leq j\leq n}|x_j| = \|\mathbf{x}\|_\infty$$

and hence

$$\|\mathbf{x}\|_1 = \sum_{i=1}^{n}|x_i| \leq n\|\mathbf{x}\|_\infty$$

(b) $$\|\mathbf{x}\|_2 = \left(\sum_{i=1}^{n} x_i^2\right)^{1/2} \leq \left(\sum_{i=1}^{n}(\max_{1\leq j\leq n}|x_j|)^2\right)^{1/2}$$
$$= (n(\max_{1\leq j\leq n}|x_j|^2))^{1/2} = \sqrt{n}\|\mathbf{x}\|_\infty$$

If $\mathbf{x}$ is a vector whose entries are all equal to 1 then for this vector equality will hold in parts (a) and (b) since

$$\|\mathbf{x}\|_\infty = 1, \quad \|\mathbf{x}\|_1 = n, \quad \|\mathbf{x}\|_2 = \sqrt{n}$$

28. Each norm produces a different unit "circle".

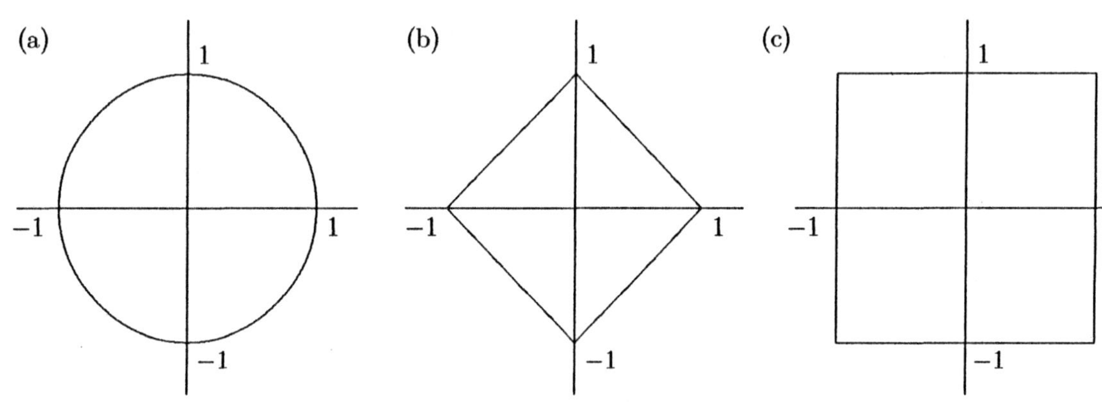

29. (a) $\langle A\mathbf{x}, \mathbf{y}\rangle = (A\mathbf{x})^T\mathbf{y} = \mathbf{x}^T A^T\mathbf{y} = \langle \mathbf{x}, A^T\mathbf{y}\rangle$
(b) $\langle A^TA\mathbf{x}, \mathbf{x}\rangle = \langle \mathbf{x}, A^TA\mathbf{x}\rangle = \mathbf{x}^TA^TA\mathbf{x} = (A\mathbf{x})^TA\mathbf{x} = \langle A\mathbf{x}, A\mathbf{x}\rangle = \|A\mathbf{x}\|^2$

SECTION 5

2. (a) $\mathbf{u}_1^T\mathbf{u}_1 = \frac{1}{18} + \frac{1}{18} + \frac{16}{18} = 1$

$\mathbf{u}_2^T\mathbf{u}_2 = \frac{4}{9} + \frac{4}{9} + \frac{1}{9} = 1$

$$\mathbf{u}_3^T\mathbf{u}_3 = \frac{1}{2} + \frac{1}{2} + 0 = 1$$

$$\mathbf{u}_1^T\mathbf{u}_2 = \frac{\sqrt{2}}{9} + \frac{\sqrt{2}}{9} - \frac{2\sqrt{2}}{9} = 0$$

$$\mathbf{u}_1^T\mathbf{u}_3 = \frac{1}{6} - \frac{1}{6} + 0 = 0$$

$$\mathbf{u}_2^T\mathbf{u}_3 = \frac{\sqrt{2}}{3} - \frac{\sqrt{2}}{3} + 0 = 0$$

4. (a) $\mathbf{x}_1^T\mathbf{x}_1 = \cos^2\theta + \sin^2\theta = 1$

$\mathbf{x}_2^T\mathbf{x}_2 = (-\sin\theta)^2 + \cos^2\theta = 1$

$\mathbf{x}_1^T\mathbf{x}_2 = -\cos\theta\sin\theta + \sin\theta\cos\theta = 0$

(c)
$$\begin{aligned} c_1^2 + c_2^2 &= (y_1\cos\theta + y_2\sin\theta)^2 + (-y_1\sin\theta + y_2\cos\theta)^2 \\ &= y_1^2\cos^2\theta + 2y_1y_2\sin\theta\cos\theta + y_2^2\sin^2\theta \\ &\quad + y_1^2\sin^2\theta - 2y_1y_2\sin\theta\cos\theta + y_2^2\cos^2\theta \\ &= y_1^2 + y_2^2. \end{aligned}$$

5. If $c_1 = \mathbf{u}^T\mathbf{u}_1 = \frac{1}{2}$ and $c_2 = \mathbf{u}^T\mathbf{u}_2$, then by Theorem 5.5.2

$$\mathbf{u} = c_1\mathbf{u}_1 + c_2\mathbf{u}_2$$

It follows from Parseval's formula that

$$1 = \|\mathbf{u}\|^2 = c_1^2 + c_2^2 = \frac{1}{4} + c_2^2$$

Hence

$$|\mathbf{u}^T\mathbf{u}_2| = |c_2| = \frac{\sqrt{3}}{2}$$

. 7 By Parseval's formula

$$c_1^2 + c_2^2 + c_3^2 = \|\mathbf{x}\|^2 = 25$$

It follows from Theorem 5.5.2 that

$$c_1 = \langle \mathbf{u}_1, \mathbf{x}\rangle = 4 \qquad \text{and} \qquad c_2 = \langle \mathbf{u}_2, \mathbf{x}\rangle = 0$$

Plugging these values into Parseval's formula we get

$$16 + 0 + c_3^2 = 25$$

and hence $c_3 = \pm 3$.

8. Since $\{\sin x, \cos x\}$ is an orthonormal set it follows that

$$\langle f, g\rangle = 3\cdot 1 + 2\cdot(-1) = 1$$

9. (a)
$$\begin{aligned} \sin^4 x &= \left(\frac{1-\cos 2x}{2}\right)^2 \\ &= \frac{1}{4}\cos^2 2x - \frac{1}{2}\cos 2x + \frac{1}{4} \end{aligned}$$

$$= \frac{1}{4}\left(\frac{1+\cos 4x}{2}\right) - \frac{1}{2}\cos 2x + \frac{1}{4}$$

$$= \frac{1}{8}\cos 4x - \frac{1}{2}\cos 2x + \frac{3\sqrt{2}}{8}\frac{1}{\sqrt{2}}$$

(b) (i) $\int_{-\pi}^{\pi} \sin^4 x \cos x \, dx = \pi \cdot 0 = 0$

(ii) $\int_{-\pi}^{\pi} \sin^4 x \cos 2x \, dx = \pi(-\frac{1}{2}) = -\frac{\pi}{2}$

(iii) $\int_{-\pi}^{\pi} \sin^4 x \cos 3x \, dx = \pi \cdot 0 = 0$

(iv) $\int_{-\pi}^{\pi} \sin^4 x \cos 4x \, dx = \pi \cdot \frac{1}{8} = \frac{\pi}{8}$

10. The key to seeing why $F_8 P_8$ can be partitioned into block form

$$\begin{pmatrix} F_4 & D_4 F_4 \\ F_4 & -D_4 F_4 \end{pmatrix}$$

is to note that

$$\omega_8^{2k} = e^{-\frac{4k\pi i}{8}} = e^{-\frac{2k\pi i}{4}} = \omega_4^k$$

and there are repeating patterns in the powers of ω_8. Since

$$\omega_8^4 = -1 \qquad \text{and} \qquad \omega_8^{8n} = e^{-2n\pi i} = 1$$

it follows that

$$\omega_8^{j+4} = -\omega_8^j \qquad \text{and} \qquad \omega_8^{8n+j} = \omega_8^j$$

Using these results let us examine the odd and even columns of F_8. Let us denote the jth column vector of the $m \times m$ Fourier matrix by $\mathbf{f}_j^{(m)}$. The odd columns of the 8×8 Fourier matrix are of the form

$$\mathbf{f}_{2n+1}^{(8)} = \begin{pmatrix} \omega_8^0 \\ \omega_8^{2n} \\ \omega_8^{4n} \\ \omega_8^{6n} \\ \omega_8^{8n} \\ \omega_8^{10n} \\ \omega_8^{12n} \\ \omega_8^{14n} \end{pmatrix} = \begin{pmatrix} 1 \\ \omega_8^{2n} \\ \omega_8^{4n} \\ \omega_8^{6n} \\ 1 \\ \omega_8^{2n} \\ \omega_8^{4n} \\ \omega_8^{6n} \end{pmatrix} = \begin{pmatrix} 1 \\ \omega_4^{n} \\ \omega_4^{2n} \\ \omega_4^{3n} \\ 1 \\ \omega_4^{n} \\ \omega_4^{2n} \\ \omega_4^{3n} \end{pmatrix} = \begin{pmatrix} \mathbf{f}_{n+1}^{(4)} \\ \mathbf{f}_{n+1}^{(4)} \end{pmatrix}$$

for $n = 0, 1, 2, 3$. The even columns are of the form

$$\mathbf{f}_{2n+2}^{(8)} = \begin{pmatrix} \omega_8^0 \\ \omega_8^{2n+1} \\ \omega_8^{2(2n+1)} \\ \omega_8^{3(2n+1)} \\ \omega_8^{4(2n+1)} \\ \omega_8^{5(2n+1)} \\ \omega_8^{6(2n+1)} \\ \omega_8^{7(2n+1)} \end{pmatrix} = \begin{pmatrix} 1 \\ \omega_8\omega_8^{2n} \\ \omega_8^2\omega_8^{4n} \\ \omega_8^3\omega_8^{6n} \\ -1 \\ -\omega_8\omega_8^{2n} \\ -\omega_8^2\omega_8^{4n} \\ -\omega_8^3\omega_8^{6n} \end{pmatrix} = \begin{pmatrix} 1 \\ \omega_8\omega_4^{n} \\ \omega_8^2\omega_4^{2n} \\ \omega_8^3\omega_4^{3n} \\ -1 \\ -\omega_8\omega_4^{n} \\ -\omega_8^2\omega_4^{2n} \\ -\omega_8^3\omega_4^{3n} \end{pmatrix} = \begin{pmatrix} D_4\mathbf{f}_{n+1}^{(4)} \\ -D_4\mathbf{f}_{n+1}^{(4)} \end{pmatrix}$$

for $n = 0, 1, 2, 3$.

11. If Q is orthogonal then

$$(Q^T)^T(Q^T) = QQ^T = QQ^{-1} = I$$

Therefore Q^T is orthogonal.

12. Let θ denote the angle between $\mathbf{x}$ and $\mathbf{y}$ and let θ_1 denote the angle between $Q\mathbf{x}$ and $Q\mathbf{y}$. It follows that

$$\cos\theta_1 = \frac{(Q\mathbf{x})^TQ\mathbf{y}}{\|Q\mathbf{x}\|\|Q\mathbf{y}\|} = \frac{\mathbf{x}^TQ^TQ\mathbf{y}}{\|\mathbf{x}\|\|\mathbf{y}\|} = \frac{\mathbf{x}^T\mathbf{y}}{\|\mathbf{x}\|\|\mathbf{y}\|} = \cos\theta$$

and hence the angles are the same.

13. (a) Use mathematical induction to prove

$$(Q^m)^{-1} = (Q^T)^m = (Q^m)^T, \qquad m = 1, 2, \ldots$$

Proof: The case $m = 1$ follows from Theorem 5.5.5. If for some positive integer k

$$(Q^k)^{-1} = (Q^T)^k = (Q^k)^T$$

then

$$(Q^T)^{k+1} = Q^T(Q^T)^k = Q^T(Q^k)^T = (Q^kQ)^T = (Q^{k+1})^T$$

and

$$(Q^T)^{k+1} = Q^T(Q^T)^k = Q^{-1}(Q^k)^{-1} = (Q^kQ)^{-1} = (Q^{k+1})^{-1}$$

(b) Prove: $\|Q^m\mathbf{x}\| = \|\mathbf{x}\|$ for $m = 1, 2, \ldots$.
Proof: In the case $m = 1$

$$\|Q\mathbf{x}\|^2 = (Q\mathbf{x})^TQ\mathbf{x} = \mathbf{x}^TQ^TQ\mathbf{x} = \mathbf{x}^T\mathbf{x} = \|\mathbf{x}\|^2$$

and hence

$$\|Q\mathbf{x}\| = \|\mathbf{x}\|$$

If $\|Q^k\mathbf{y}\| = \|\mathbf{y}\|$ for any $\mathbf{y} \in R^n$, then in particular, if $\mathbf{x}$ is an arbitrary vector in R^n and we define $\mathbf{y} = Q\mathbf{x}$, then

$$\|Q^{k+1}\mathbf{x}\| = \|Q^k(Q\mathbf{x})\| = \|Q^k\mathbf{y}\| = \|\mathbf{y}\| = \|Q\mathbf{x}\| = \|\mathbf{x}\|$$

14. $H^T = (I - 2\mathbf{u}\mathbf{u}^T)^T = I^T - 2(\mathbf{u}^T)^T\mathbf{u}^T = I - 2\mathbf{u}\mathbf{u}^T = H$

$$\begin{aligned} H^TH &= H^2 \\ &= (I - 2\mathbf{u}\mathbf{u}^T)^2 \\ &= I - 4\mathbf{u}\mathbf{u}^T + 4\mathbf{u}\mathbf{u}^T\mathbf{u}\mathbf{u}^T \\ &= I - 4\mathbf{u}\mathbf{u}^T + 4\mathbf{u}\mathbf{u}^T \\ &= I \end{aligned}$$

15. Since $Q^TQ = I$, it follows that

$$[\det(Q)]^2 = \det(Q^T)\det(Q) = \det(I) = 1$$

Thus $\det(Q) = \pm 1$.

16. (a) Let Q_1 and Q_2 be orthogonal $n \times n$ matrices and let $Q = Q_1Q_2$. It follows that

$$Q^TQ = (Q_1Q_2)^TQ_1Q_2 = Q_2^TQ_1^TQ_1Q_2 = I$$

Therefore Q is orthogonal.

(b) Yes. Let P_1 and P_2 be permutation matrices. The columns of P_1 are the same as the columns of I, but in a different order. Postmultiplication of P_1 by P_2 reorders the columns of P_1. Thus P_1P_2 is a matrix formed by reordering the columns of I and hence is a permutation matrix.

17. There are $n!$ permutations of any set with n distinct elements. Therefore there are $n!$ possible permutations of the row vectors of the $n \times n$ identity matrix and hence the number of $n \times n$ permutation matrices is $n!$.

18. A permutation P is an orthogonal matrix so $P^T = P^{-1}$ and if P is a symmetric permutation matrix then $P = P^T = P^{-1}$ and hence

$$P^2 = P^TP = P^{-1}P = I$$

So for a symmetric permutation matrix we have

$$P^{2k} = (P^2)^k = I^k = I \qquad \text{and} \qquad P^{2k+1} = PP^{2k} = PI = P$$

19.

$$\begin{aligned} I = UU^T &= (\mathbf{u}_1, \mathbf{u}_2, \ldots, \mathbf{u}_n)\begin{pmatrix} \mathbf{u}_1^T \\ \mathbf{u}_2^T \\ \vdots \\ \mathbf{u}_n^T \end{pmatrix} \\ &= \mathbf{u}_1\mathbf{u}_1^T + \mathbf{u}_2\mathbf{u}_2^T + \cdots + \mathbf{u}_n\mathbf{u}_n^T \end{aligned}$$

20. The proof is by induction on n. If $n = 1$, then Q must be either (1) or (-1). Assume the result holds for all $k \times k$ upper triangular orthogonal matrices and let Q be a $(k+1)\times(k+1)$ matrix that is upper triangular and orthogonal. Since Q is upper triangular its first column must be a multiple of $\mathbf{e}_1$. But Q

is also orthogonal, so $\mathbf{q}_1$ is a unit vector. Thus $\mathbf{q}_1 = \pm\mathbf{e}_1$. Furthermore, for $j = 2, \ldots, n$

$$q_{1j} = \mathbf{e}_1^T \mathbf{q}_j = \pm\mathbf{q}_1^T \mathbf{q}_j = 0$$

Thus Q must be of the form

$$Q = \begin{pmatrix} \pm 1 & 0 & 0 & \cdots & 0 \\ \mathbf{0} & \mathbf{p}_2 & \mathbf{p}_3 & \cdots & \mathbf{p}_{k+1} \end{pmatrix}$$

The matrix $P = (\mathbf{p}_2, \mathbf{p}_3, \ldots, \mathbf{p}_{k+1})$ is a $k \times k$ matrix that is both upper triangular and orthogonal. By the induction hypothesis P must be a diagonal matrix with diagonal entries equal to ± 1. Thus Q must also be a diagonal matrix with ± 1's on the diagonal.

21. (a) The columns of A form an orthonormal set since

$$\begin{aligned} \mathbf{a}_1^T \mathbf{a}_2 &= -\frac{1}{4} - \frac{1}{4} + \frac{1}{4} + \frac{1}{4} = 0 \\ \mathbf{a}_1^T \mathbf{a}_1 &= \frac{1}{4} + \frac{1}{4} + \frac{1}{4} + \frac{1}{4} = 1 \\ \mathbf{a}_2^T \mathbf{a}_2 &= \frac{1}{4} + \frac{1}{4} + \frac{1}{4} + \frac{1}{4} = 1 \end{aligned}$$

22. (b)
(i) $A\mathbf{x} = P\mathbf{b} = (2, 2, 0, 0)^T$
(ii) $A\mathbf{x} = P\mathbf{b} = \left(\frac{3}{2}, \frac{3}{2}, \frac{7}{2}, \frac{7}{2}\right)^T$
(iii) $A\mathbf{x} = P\mathbf{b} = (1, 1, 2, 2)^T$

23. (a) One can find a basis for $N(A^T)$ in the usual way by computing the reduced row echelon form of A^T.

$$\begin{pmatrix} \frac{1}{2} & \frac{1}{2} & \frac{1}{2} & \frac{1}{2} \\ -\frac{1}{2} & -\frac{1}{2} & \frac{1}{2} & \frac{1}{2} \end{pmatrix} \rightarrow \begin{pmatrix} 1 & 1 & 0 & 0 \\ 0 & 0 & 1 & 1 \end{pmatrix}$$

Setting the free variables equal to one and solving for the lead variables, we end up with basis vectors $\mathbf{x}_1 = (-1, 1, 0, 0)^T$, $\mathbf{x} = (0, 0, -1, 1)^T$. Since these vectors are already orthogonal we need only normalize to obtain an orthonormal basis for $N(A^T)$.

$$\mathbf{u}_1 = \frac{1}{\sqrt{2}}(-1, 1, 0, 0)^T \qquad \mathbf{u}_2 = \frac{1}{\sqrt{2}}(0, 0, -1, 1)^T$$

24. (a) Let U_1 be a matrix whose columns form an orthonormal basis for $R(A)$ and let U_2 be a matrix whose columns form an orthonormal basis for $N(A^T)$. If we set $U = (U_1, U_2)$, then since $R(A)$ and $N(A^T)$ are orthogonal complements in R^n, it follows that U is an orthogonal matrix. The unique projection matrix P onto $R(A)$ is given $P = U_1 U_1^T$ and the projection matrix onto $N(A^T)$ is given by $U_2 U_2^T$. Since U is orthogonal

it follows that

$$I = UU^T = U_1U_1^T + U_2U_2^T = P + U_2U_2^T$$

Thus the projection matrix onto $N(A^T)$ is given by

$$U_2U_2^T = I - P$$

(b) The proof here is essentially the same as in part (a). Let V_1 be a matrix whose columns form an orthonormal basis for $R(A^T)$ and let V_2 be a matrix whose columns form an orthonormal basis for $N(A)$. If we set $V = (V_1, V_2)$, then since $R(A^T)$ and $N(A)$ are orthogonal complements in R^m, it follows that V is an orthogonal matrix. The unique projection matrix Q onto $R(A^T)$ is given $Q = V_1V_1^T$ and the projection matrix onto $N(A)$ is given by $V_2V_2^T$. Since V is orthogonal it follows that

$$I = VV^T = V_1V_1^T + V_2V_2^T = Q + V_2V_2^T$$

Thus the projection matrix onto $N(A)$ is given by

$$V_2V_2^T = I - Q$$

25. (a) If U is a matrix whose columns form an orthonormal basis for S, then the projection matrix P corresponding to S is given by $P = UU^T$. It follow that

$$P^2 = (UU^T)(UU^T) = U(U^TU)U^T = UIU^T = P$$

(b) $P^T = (UU^T)^T = (U^T)^TU^T = UU^T = P$

26. The (i, j) entry of A^TA will be $\mathbf{a}_i^T\mathbf{a}_j$. This will be 0 if $i \neq j$. Thus A^TA is a diagonal matrix with diagonal elements $\mathbf{a}_1^T\mathbf{a}_1, \mathbf{a}_2^T\mathbf{a}_2, \ldots, \mathbf{a}_n^T\mathbf{a}_n$. The ith entry of $A^T\mathbf{b}$ is $\mathbf{a}_i^T\mathbf{b}$. Thus if $\hat{\mathbf{x}}$ is the solution to the normal equations, its ith entry will be

$$\hat{\mathbf{x}}_i = \frac{\mathbf{a}_i^T\mathbf{b}}{\mathbf{a}_i^T\mathbf{a}_i} = \frac{\mathbf{b}^T\mathbf{a}_i}{\mathbf{a}_i^T\mathbf{a}_i}$$

27. (a) $\langle 1, x\rangle = \int_{-1}^{1} 1x\,dx = \left.\frac{x^2}{2}\right|_{-1}^{1} = 0$

28. (a) $\langle 1, 2x-1\rangle = \int_0^1 1\cdot(2x-1)dx = \left. x^2 - x\right|_0^1 = 0$

(b) $\|1\|^2 = \langle 1, 1\rangle = \int_0^1 1\cdot 1\,dx = \left. x\right|_0^1 = 1$

$$\|2x-1\|^2 = \int_0^1 (2x-1)^2dx = \frac{1}{3}$$

Therefore

$$\|1\| = 1 \qquad \text{and} \qquad \|2x-1\| = \frac{1}{\sqrt{3}}$$

(c) The best least squares approximation to $\sqrt{x}$ from S is given by

$$\ell(x) = c_1 1 + c_2\sqrt{3}(2x-1)$$

where

$$\begin{aligned} c_1 &= \langle 1, x^{1/2} \rangle = \int_0^1 1\, x^{1/2} dx = \frac{2}{3} \\ c_2 &= \langle \sqrt{3}(2x-1), x^{1/2} \rangle = \int_0^1 \sqrt{3}(2x-1)x^{1/2} dx = \frac{2\sqrt{3}}{15} \end{aligned}$$

Thus

$$\begin{aligned} \ell(x) &= \frac{2}{3} \cdot 1 + \frac{2\sqrt{3}}{15}(\sqrt{3}(2x-1)) \\ &= \frac{4}{5}(x + \frac{1}{3}) \end{aligned}$$

29. We saw in Example 3 that $\{1/\sqrt{2}, \cos x, \cos 2x, \ldots, \cos nx\}$ is an orthonormal set. In Section 4, Exercise 9 we saw that the functions $\cos kx$ and $\sin jx$ were orthogonal unit vectors in $C[-\pi, \pi]$. Furthermore

$$\left\langle \frac{1}{\sqrt{2}}, \sin jx \right\rangle = \frac{1}{\pi} \int_{-\pi}^{\pi} \frac{1}{\sqrt{2}} \sin jx\ dx = 0$$

Therefore $\{1/\sqrt{2}, \cos x, \cos 2x, \ldots, \cos nx, \sin x, \sin 2x, \ldots, \sin nx\}$ is an orthonormal set of vectors.

30. The coefficients of the best approximation are given by

$$a_0 = \langle 1, |x| \rangle = \frac{1}{\pi} \int_{-\pi}^{\pi} 1 \cdot |x|\, dx = \frac{2}{\pi} \int_0^{\pi} x\, dx = \pi$$

$$a_1 = \langle \cos x, |x| \rangle = \frac{2}{\pi} \int_0^{\pi} x \cos x\, dx = -\frac{4}{\pi}$$

$$a_2 = \frac{2}{\pi} \int_0^{\pi} x \cos 2x\, dx = 0$$

To compute the coefficients of the sin terms we must integrate $x \sin x$ and $x \sin 2x$ from $-\pi$ to π. Since both of these are odd functions the integrals will be 0. Therefore $b_1 = b_2 = 0$. The best trigonometric approximation of degree 2 or less is given by

$$p(x) = \frac{\pi}{2} - \frac{4}{\pi} \cos x$$

31. If $\mathbf{u} = c_1\mathbf{x}_1 + c_2\mathbf{x}_2 + \cdots + c_k\mathbf{x}_k$ is an element of S_1 and $\mathbf{v} = c_{k+1}\mathbf{x}_{k+1} + c_{k+2}\mathbf{x}_{k+2} + \cdots + c_n\mathbf{x}_n$ is an element of S_2, then

$$\begin{aligned} \langle \mathbf{u}, \mathbf{v} \rangle &= \left\langle \sum_{i=1}^{k} c_i\mathbf{x}_i, \sum_{j=k+1}^{n} c_j\mathbf{x}_j \right\rangle \\ &= \sum_{k=1}^{k} \sum_{j=k+1}^{n} c_i c_j \langle \mathbf{x}_i, \mathbf{x}_j \rangle \\ &= 0 \end{aligned}$$

32. (a) By Theorem 5.5.2,

$$\begin{aligned}\mathbf{x} &= \sum_{i=1}^{n}\langle \mathbf{x}, \mathbf{x}_i\rangle \mathbf{x}_i \\ &= \sum_{i=1}^{k}\langle \mathbf{x}, \mathbf{x}_i\rangle \mathbf{x}_i + \sum_{i=k+1}^{n}\langle \mathbf{x}, \mathbf{x}_i\rangle \mathbf{x}_i \\ &= \mathbf{p}_1 + \mathbf{p}_2\end{aligned}$$

(b) It follows from Exercise 31 that $S_2 \subset S_1^{\perp}$. On the other hand if $\mathbf{x} \in S_1^{\perp}$ then by part (a) $\mathbf{x} = \mathbf{p}_1 + \mathbf{p}_2$. Since $\mathbf{x} \in S_1^{\perp}$, $\langle \mathbf{x}, \mathbf{x}_i\rangle = 0$ for $i = 1, \ldots, k$. Thus $\mathbf{p}_1 = \mathbf{0}$ and $\mathbf{x} = \mathbf{p}_2 \in S_2$. Therefore $S_2 = S^{\perp}$.

33. Let

$$\mathbf{u}_i = \frac{1}{\|\mathbf{x}_i\|}\mathbf{x}_i \quad \text{for} \quad i = 1, \ldots, n$$

By Theorem 5.5.8 the best least squares approximation to $\mathbf{x}$ from S is given by

$$\begin{aligned}\mathbf{p} = \sum_{i=1}^{n}\langle \mathbf{x}, \mathbf{u}_i\rangle \mathbf{u}_i &= \sum_{i=1}^{n}\frac{1}{\|\mathbf{x}_i\|^2}\langle \mathbf{x}, \mathbf{x}_i\rangle \mathbf{x}_i \\ &= \sum_{i=1}^{n}\frac{\langle \mathbf{x}, \mathbf{x}_i\rangle}{\langle \mathbf{x}_i, \mathbf{x}_i\rangle}\mathbf{x}_i.\end{aligned}$$

SECTION 6

9. $r_{11} = \|\mathbf{x}_1\| = 5$

$$\mathbf{q}_1 = \frac{1}{r_{11}}\mathbf{x}_1 = \left(\frac{4}{5}, \frac{2}{5}, \frac{2}{5}, \frac{1}{5}\right)^T$$

$r_{12} = \mathbf{q}_1^T\mathbf{x}_2 = 2$ and $r_{13} = \mathbf{q}_1^T\mathbf{x}_3 = 1$

$$\mathbf{x}_2^{(1)} = \mathbf{x}_2 - r_{12}\mathbf{q}_1 = \left(\frac{2}{5}, -\frac{4}{5}, -\frac{4}{5}, \frac{8}{5}\right)^T, \mathbf{x}_3^{(1)} = \mathbf{x}_3 - r_{13}\mathbf{q}_1 = \left(\frac{1}{5}, \frac{3}{5}, -\frac{7}{5}, \frac{4}{5}\right)^T$$

$r_{22} = \|\mathbf{x}_2^{(1)}\| = 2$

$$\mathbf{q}_2 = \frac{1}{r_{22}}\mathbf{x}_2^{(1)} = \left(\frac{1}{5}, -\frac{2}{5}, -\frac{2}{5}, \frac{4}{5}\right)^T$$

$r_{23} = \mathbf{x}_3^T\mathbf{q}_2 = 1$

$\mathbf{x}_3^{(2)} = \mathbf{x}_3^{(1)} - r_{23}\mathbf{q}_2 = (0,\ 1,\ -1,\ 0)^T$

$r_{33} = \|\mathbf{x}_3^{(2)}\| = \sqrt{2}$

$$\mathbf{q}_3 = \frac{1}{r_{33}}\mathbf{x}_3^{(2)} = \left(0, \frac{1}{\sqrt{2}}, -\frac{1}{\sqrt{2}}, 0\right)^T$$

10. Given a basis $\{x_1, \ldots, x_n\}$, one can construct an orthonormal basis using either the classical Gram–Schmidt process or the modified process. When carried out in exact arithmetic both methods will produce the same orthonormal set $\{\mathbf{q}_1, \ldots, \mathbf{q}_n\}$.

Proof: The proof is by induction on n. In the case $n = 1$, the vector $\mathbf{q}_1$ is computed in the same way for both methods.

$$\mathbf{q}_1 = \frac{1}{r_{11}}\mathbf{x}_1 \quad \text{where} \quad r_{11} = \|\mathbf{x}\|_1$$

Assume $\mathbf{q}_1, \ldots, \mathbf{q}_k$ are the same for both methods. In the classical Gram–Schmidt process one computes $\mathbf{q}_{k+1}$ as follows: Set

$$\begin{aligned}
r_{i,k+1} &= \langle \mathbf{x}_{k+1}, \mathbf{q}_i \rangle, \qquad i = 1, \ldots, k \\
\mathbf{p}_k &= r_{1,k+1}\mathbf{q}_1 + r_{2,k+1}\mathbf{q}_2 + \cdots + r_{k,k+1}\mathbf{q}_k \\
r_{k+1,k+1} &= \|\mathbf{x}_{k+1} - \mathbf{p}_k\| \\
\mathbf{q}_{k+1} &= \frac{1}{r_{k+1,k+1}}(\mathbf{x}_{k+1} - \mathbf{p}_k)
\end{aligned}$$

Thus

$$\mathbf{q}_{k+1} = \frac{1}{r_{k+1,k+1}}(\mathbf{x}_{k+1} - r_{1,k+1}\mathbf{q}_1 - r_{2,k+1}\mathbf{q}_2 - \cdots - r_{k,k+1}\mathbf{q}_k)$$

In the modified version, at step 1 the vector $r_{1,k+1}\mathbf{q}_1$ is subtracted from $\mathbf{x}_{k+1}$.

$$\mathbf{x}_{k+1}^{(1)} = \mathbf{x}_{k+1} - r_{1,k+1}\mathbf{q}_1$$

At the next step $r_{2,k+1}\mathbf{q}_2$ is subtracted from $\mathbf{x}_{k+1}^{(1)}$.

$$\begin{aligned}
\mathbf{x}_{k+1}^{(2)} &= \mathbf{x}_{k+1}^{(1)} - r_{2,k+1}\mathbf{q}_2 \\
&= \mathbf{x}_{k+1} - r_{1,k+1}\mathbf{q}_1 - r_{2,k+1}\mathbf{q}_2
\end{aligned}$$

In general after k steps we have

$$\begin{aligned}
\mathbf{x}_{k+1}^{(k)} &= \mathbf{x}_{k+1} - r_{1,k+1}\mathbf{q}_1 - r_{2,k+1}\mathbf{q}_2 - \cdots - r_{k,k+1}\mathbf{q}_k \\
&= \mathbf{x}_{k+1} - \mathbf{p}_k
\end{aligned}$$

In the last step we set

$$r_{k+1,k+1} = \|\mathbf{x}_{k+1}^{(k)}\| = \|\mathbf{x}_{k+1} - \mathbf{p}_k\|$$

and set

$$\mathbf{q}_{k+1} = \frac{1}{r_{k+1,k+1}}\mathbf{x}_{k+1}^{(k)} = \frac{1}{r_{k+1,k+1}}(\mathbf{x}_{k+1} - \mathbf{p}_k)$$

Thus $\mathbf{q}_{k+1}$ is the same as in the classical Gram–Schmidt process.

11. If the Gram-Schmidt process is applied to a set $\{\mathbf{v}_1, \mathbf{v}_2, \mathbf{v}_3\}$ and $\mathbf{v}_3$ is in $\text{Span}(\mathbf{v}_1, \mathbf{v}_2)$, then the process will break down at the third step. If $\mathbf{u}_1$, $\mathbf{u}_2$ have been constructed so that they form an orthonormal basis for $S_2 = \text{Span}(\mathbf{v}_1, \mathbf{v}_2)$, then the projection $\mathbf{p}_2$ of $\mathbf{v}_3$ onto S_2 is $\mathbf{v}_3$ (since $\mathbf{v}_3$ is already in S_2). Thus $\mathbf{v}_3 - \mathbf{p}_2$ will be the zero vector and hence we cannot normalize to obtain a unit vector $\mathbf{u}_3$.

12. (a) Since

$$\mathbf{p} = c_1\mathbf{q}_1 + c_2\mathbf{q}_2 + \cdots + c_n\mathbf{q}_n$$

is the projection of $\mathbf{b}$ onto $R(A)$ and $\mathbf{q}_1, \mathbf{q}_2, \ldots, \mathbf{q}_n$ form an orthonormal basis for $R(A)$, it follows that

$$c_j = \mathbf{q}_j^T\mathbf{b} \qquad j = 1, \ldots, n$$

and hence

$$\mathbf{c} = Q^T\mathbf{b}$$

(b) $\mathbf{p} = c_1\mathbf{q}_1 + c_2\mathbf{q}_2 + \cdots + c_n\mathbf{q}_n = Q\mathbf{c} = QQ^T\mathbf{b}$

(c) Both $A(A^TA)^{-1}A^T$ and QQ^T are projection matrices that project vectors onto $R(A)$. Since the projection matrix is unique for a given subspace it follows that

$$QQ^T = A(A^TA)^{-1}A^T$$

13. **(a)** If $\{\mathbf{v}_1, \ldots, \mathbf{v}_k\}$ is an orthonormal basis for V then by Theorem 3.4.4 it can be extended to form a basis $\{\mathbf{v}_1, \ldots, \mathbf{v}_k, \mathbf{u}_{k+1}, \mathbf{u}_{k+2}, \ldots, \mathbf{u}_m\}$ for U. If we apply the Gram-Schmidt process to this basis, then since $\mathbf{v}_1, \ldots, \mathbf{v}_k$ are already orthonormal vectors, they will remain unchanged and we with end up with an orthonormal basis $\{\mathbf{v}_1, \ldots, \mathbf{v}_k, \mathbf{v}_{k+1}, \ldots, \mathbf{v}_m\}$.

(b) If $\mathbf{u}$ is any vector in U, we can write

(3) $\quad \mathbf{u} = c_1\mathbf{v}_1 + \cdots + c_k\mathbf{v}_k + c_{k+1}\mathbf{v}_{k+1} + \cdots + c_m\mathbf{v}_m = \mathbf{v} + \mathbf{w}$

where

$$\mathbf{v} = c_1\mathbf{v}_1 + \cdots + c_k\mathbf{v}_k \in V \qquad \text{and} \qquad \mathbf{w} = c_{k+1}\mathbf{v}_{k+1} + \cdots + c_m\mathbf{v}_m) \in W$$

Therefore, $U = V + W$. The representation (3) is unique. Indeed if

$$\mathbf{u} = \mathbf{v} + \mathbf{w} = \mathbf{x} + \mathbf{y}$$

where $\mathbf{v}, \mathbf{x}$ are in V and $\mathbf{w}, \mathbf{y}$ are in W, then

$$\mathbf{v} - \mathbf{x} = \mathbf{y} - \mathbf{w}$$

and hence $\mathbf{v} - \mathbf{x} \in V \cap W$. Since V and W are orthogonal subspaces we have $V \cap W = \{\mathbf{0}\}$ and hence $\mathbf{v} = \mathbf{x}$. By the same reasoning $\mathbf{w} = \mathbf{y}$. It follows then that $U = V \oplus W$.

14. Let $m = \dim U$, $k = \dim V$, and $W = U \cap V$. If $\dim W = r > 0$ and $\{\mathbf{v}_1, \ldots, \mathbf{v}_r\}$ is a basis for W, then by Exercise 13(a) we can extend this basis to an orthonormal basis $\{\mathbf{v}_1, \ldots, \mathbf{v}_r, \mathbf{v}_{r+1}, \ldots, \mathbf{v}_k\}$ for V. Let

$$V_1 = \text{Span}(\mathbf{v}_{r+1}, \ldots, \mathbf{v}_k)$$

By Exercise 13(b) we have $V = W \oplus V_1$. We claim that $U + V = U \oplus V_1$. Since V_1 is a subspace of V it follows that $U + V_1$ is a subspace of $U + V$. On the other hand, if $\mathbf{x}$ is in $U + V$ then

$$\mathbf{x} = \mathbf{u} + \mathbf{v} = \mathbf{u} + (\mathbf{w} + \mathbf{v}_1) = (\mathbf{u} + \mathbf{w}) + \mathbf{v}_1$$

where $\mathbf{u} \in U$, $\mathbf{v} \in V$, $\mathbf{w} \in W$, and $\mathbf{v}_1 \in V_1$. Since $\mathbf{u} + \mathbf{w}$ is in U it follows that $\mathbf{x}$ is in $U + V_1$ and hence $U + V = U + V_1$. To show that we have a direct sum we must show that $U \cap V_1 = \{\mathbf{0}\}$. If $\mathbf{z} \in U \cap V_1$ then $\mathbf{z}$ is also in the larger subspace $W = U \cap V$. So $\mathbf{z}$ is in both V_1 and W. However, by construction V_1 is orthogonal to W, so the intersection of the two subspaces must be $\{\mathbf{0}\}$. Therefore $U \cap V_1 = \{\mathbf{0}\}$. It follows then that

$$U + V = U \oplus V$$

and hence

$$\begin{aligned}\dim(U+V) &= \dim(U \oplus V) = \dim U + \dim V_1 \\ &= m + (k-r) = m + k - r \\ &= \dim U + \dim V - \dim(U \cap V)\end{aligned}$$

SECTION 7

3. Let $x = \cos\theta$.

(a) $$\begin{aligned}2T_m(x)T_n(x) &= 2\cos m\theta \cos n\theta \\ &= \cos(m+n)\theta + \cos(m-n)\theta \\ &= T_{m+n}(x) + T_{m-n}(x)\end{aligned}$$

(b) $T_m(T_n(x)) = T_m(\cos n\theta) = \cos(mn\theta) = T_{mn}(x)$

5. $p_n(x) = a_n x^n + q(x)$ where degree $q(x) < n$. By Theorem 5.7.1, $\langle q, p_n\rangle = 0$. It follows then that

$$\begin{aligned}\|p_n\|^2 &= \langle a_n x^n + q(x), p(x)\rangle \\ &= a_n\langle x^n, p_n\rangle + \langle q, p_n\rangle \\ &= a_n\langle x^n, p_n\rangle\end{aligned}$$

6. (b) $$\begin{aligned}U_{n-1}(x) &= \frac{1}{n}T_n'(x) \\ &= \frac{1}{n}\frac{dT_n}{d\theta}\Big/\frac{dx}{d\theta} \\ &= \frac{\sin n\theta}{\sin\theta}\end{aligned}$$

7. (a) $$\begin{aligned}U_n(x) - xU_{n-1}(x) &= \frac{\sin(n+1)\theta}{\sin\theta} - \frac{\cos\theta\sin n\theta}{\sin\theta} \\ &= \frac{\sin n\theta\cos\theta + \cos n\theta\sin\theta - \cos\theta\sin n\theta}{\sin\theta} \\ &= \cos n\theta \\ &= T_n(x)\end{aligned}$$

(b) $$\begin{aligned}U_n(x) + U_{n-2}(x) &= \frac{\sin(n+1)\theta + \sin(n-1)\theta}{\sin\theta} \\ &= \frac{2\sin n\theta\cos\theta}{\sin\theta} \\ &= 2xU_{n-1}(x)\end{aligned}$$

$$U_n(x) = 2xU_{n-1}(x) - U_{n-2}(x)$$

8. $$\langle U_n, U_m\rangle = \int_{-1}^{1} U_n(x)U_m(x)(1-x^2)^{1/2}dx$$
$$= \int_0^{\pi} \sin[(n+1)\theta]\sin[(m+1)\theta]d\theta \quad (x=\cos\theta)$$
$$= 0 \text{ if } m \neq n$$

9. (i) $n=0,\ y=1,\ y'=0,\ y''=0$
$(1-x^2)y'' - 2xy' + 0\cdot 1\cdot 1 = 0$

(ii) $n=1,\ y=P_1(x)=x,\ y'=1,\ y''=0$
$(1-x^2)\cdot 0 - 2x\cdot 1 + 1\cdot 2x = 0$

(iii) $n=2,\ y=P_2(x)=\frac{3}{2}\left(x^2-\frac{1}{3}\right),\ y'=3x,\ y''=3$
$(1-x^2)\cdot 3 - 2x\cdot 3x + 6\cdot\frac{3}{2}\left(x^2-\frac{1}{3}\right) = 0$

10. (a) Prove: $H_n'(x) = 2nH_{n-1}(x),\ n=0,1,2,\ldots$.
Proof: The proof is by mathematical induction. In the case $n=0$

$$H_0'(x) = 0 = 2nH_{-1}(x)$$

Assume

$$H_k'(x) = 2kH_{k-1}(x)$$

for all $k \leq n$.

$$H_{n+1}(x) = 2xH_n(x) - 2nH_{n-1}(x)$$

Differentiating both sides we get

$$\begin{aligned} H_{n+1}'(x) &= 2H_n + 2xH_n' - 2nH_{n-1}' \\ &= 2H_n + 2x[2nH_{n-1}] - 2n[2(n-1)H_{n-2}] \\ &= 2H_n + 2n[2xH_{n-1} - 2(n-1)H_{n-2}] \\ &= 2H_n + 2nH_n \\ &= 2(n+1)H_n \end{aligned}$$

(b) Prove: $H_n''(x) - 2xH_n'(x) + 2nH_n(x) = 0,\ n=0,1,\ldots$.
Proof: It follows from part (a) that

$$\begin{aligned} H_n'(x) &= 2nH_{n-1}(x) \\ H_n''(x) &= 2nH_{n-1}'(x) = 4n(n-1)H_{n-2}(x) \end{aligned}$$

Therefore

$$\begin{aligned} &H_n''(x) - 2xH_n'(x) + 2nH_n(x) \\ &\quad = 4n(n-1)H_{n-2}(x) - 4xnH_{n-1}(x) + 2nH_n(x) \\ &\quad = 2n[H_n(x) - 2xH_{n-1}(x) + 2(n-1)H_{n-2}(x)] \\ &\quad = 0 \end{aligned}$$

12. If $f(x)$ is a polynomial of degree less than n and $P(x)$ is the Lagrange interpolating polynomial that agrees with $f(x)$ at $x_1, \ldots, x_n$, then degree $P(x) \leq n-1$. If we set

$$h(x) = P(x) - f(x)$$

then the degree of h is also $\leq n-1$ and

$$h(x_i) = P(x_i) - f(x_i) = 0 \qquad i = 1, \ldots, n$$

Therefore h must be the zero polynomial and hence

$$P(x) = f(x)$$

15. (a) The quadrature formula approximates the integral of $f(x)$ by a sum which is equal to the exact value of the integral of Lagrange polynomial that interpolates f at the given points. In the case where f is a polynomial of degree less than n, the Lagrange polynomial will be equal to f, so the quadrature formula will yield the exact answer.

(b) If we take the constant function $f(x) = 1$ and apply the quadrature formula we get

$$\begin{aligned} \int_{-1}^{1} f(x)dx &= A_1 f(x_1) + A_2 f(x_2) + \cdots + A_n f(x_n) \\ \int_{-1}^{1} 1dx &= A_1 \cdot 1 + A_2 \cdot 1 + \cdots + A_n \cdot 1 \\ 2 &= A_1 + A_2 + \cdot + A_n \end{aligned}$$

16. (a) If $j \geq 1$ then the Legendre polynomial P_j is orthogonal to $P_0 = 1$. Thus we have

$$(4) \qquad \int_{-1}^{1} P_j(x)dx = \int_{-1}^{1} P_j(x)P_0(x)dx = \langle P_j, P_0 \rangle = 0 \quad (j \geq 1)$$

The n-point Gauss-Legendre quadrature formula will yield the exact value of the integral of $f(x)$ whenever $f(x)$ is a polynomial of degree less than $2n$. So in particular for $f(x) = P_j(x)$ we have

$$(5) \int_{-1}^{1} P_j(x)dx = P_j(x_1)A_1 + P_j(x_2)A_2 + \cdots + P_j(x_n)A_n \quad (0 \leq j < 2n)$$

It follows from (4) and (5) that

$$P_j(x_1)A_1 + P_j(x_2)A_2 + \cdots + P_j(x_n)A_n = 0 \quad \text{for} \ \ 1 \leq j < 2n$$

(b)

$$\begin{aligned} A_1 + A_2 + \cdots + A_n &= 2 \\ P_1(x_1)A_1 + P_1(x_2)A_2 + \cdots + P_1(x_n)A_n &= 0 \\ &\vdots \\ P_{n-1}(x_1)A_1 + P_{n-1}(x_2)A_2 + \cdots + P_{n-1}(x_n)A_n &= 0 \end{aligned}$$

17. **(a)** If $\|Q_j\| = 1$ for each j, then in the recursion relation we will have

$$\gamma_k = \frac{\langle Q_k, Q_k\rangle}{\langle Q_{k-1}, Q_{k-1}\rangle} = 1 \quad (k \geq 1)$$

and hence the recursion relation for the orthonormal sequence simplifies to

$$\alpha_{k+1}Q_{k+1}(x) = (x - \beta_{k+1}Q_k(x) - \alpha_k Q_{k-1}(x) \quad (k \geq 0)$$

where Q_{-1} is taken to be the zero polynomial.

(b) For $k = 0, \ldots, n-1$ we can rewrite the recursion relation in part (a) in the form

$$\alpha_k Q_{k-1}(x) + \beta_{k+1}Q_k(x) + \alpha_{k+1}Q_{k+1}(x) = xQ_k(x)$$

Let λ be any root of Q_n and let us plug it into each of the n-equations. Note that the first equation ($k = 0$) will be

$$\beta_1 Q_0(\lambda) + \alpha_1 Q_1(\lambda) = \lambda Q_0(\lambda)$$

since Q_{-1} is the zero polynomial. For $(2 \leq k \leq n-2)$ intermediate equations are all of the form

$$\alpha_k Q_{k-1}(\lambda) + \beta_{k+1}Q_k(\lambda) + \alpha_{k+1}Q_{k+1}(\lambda) = \lambda Q_k(\lambda)$$

The last equation ($k = n-1$) will be

$$\alpha_{n-1}Q_{n-2}(\lambda) + \beta_n Q_{n-1}(\lambda) = \lambda Q_{n-1}(\lambda)$$

since $Q_n(\lambda) = 0$. We now have a system of n equations in the variable λ. If we rewrite it in matrix form we get

$$\begin{pmatrix} \beta_1 & \alpha_1 & & & \\ \alpha_1 & \beta_2 & \alpha_2 & & \\ & \ddots & \ddots & \ddots & \\ & & \alpha_{n-2} & \beta_{n-1} & \alpha_{n-1} \\ & & & \alpha_{n-1} & \beta_n \end{pmatrix} \begin{pmatrix} Q_0(\lambda) \\ Q_1(\lambda) \\ \vdots \\ Q_{n-2}(\lambda) \\ Q_{n-1}(\lambda) \end{pmatrix} = \lambda \begin{pmatrix} Q_0(\lambda) \\ Q_1(\lambda) \\ \vdots \\ Q_{n-2}(\lambda) \\ Q_{n-1}(\lambda) \end{pmatrix}$$

MATLAB EXERCISES

1. (b) By the Cauchy-Schwarz Inequality

$$|\mathbf{x}^T\mathbf{y}| \leq \|\mathbf{x}\|\|\mathbf{y}\|$$

Therefore

$$|t| = \frac{|\mathbf{x}^T\mathbf{y}|}{\|\mathbf{x}\|\,\|\mathbf{y}\|} \leq 1$$

3. (c) From the graph it should be clear that you get a better fit at the bottom of the atmosphere.

5. (a) A is the product of two random matrices. One would expect that both of the random matrices will have full rank, that is, rank 2. Since the row vectors of A are linear combinations of the row vectors of the second random matrix, one would also expect that A would have rank 2. If the rank of A is 2, then the nullity of A should be $5 - 2 = 3$.

(b) Since the column vectors of Q form an orthonormal basis for $R(A)$ and the column vectors of W form an orthonormal basis for $N(A^T) = R(A)^{\perp}$, the column vectors of $S = (Q \;\; W)$ form an orthonormal basis for R^5 and hence S is an orthogonal matrix. Each column vector of W is in $N(A^T)$. thus it follows that

$$A^T W = O$$

and

$$W^T A = (A^T W)^T = O^T$$

(c) Since S is an orthogonal matrix, we have

$$I = SS^T = (Q \;\; W)\begin{pmatrix} Q^T \\ W^T \end{pmatrix} = QQ^T + WW^T$$

Thus

$$QQ^T = I - WW^T$$

and it follows that

$$QQ^T A = A - WW^T A = A - WO = A$$

(d) If $\mathbf{b} \in R(A)$, then $\mathbf{b} = A\mathbf{x}$ for some $\mathbf{x} \in R^5$. It follows from part (c) that

$$QQ^T\mathbf{b} = QQ^T(A\mathbf{x}) = (QQ^T A)\mathbf{x} = A\mathbf{x} = \mathbf{b}$$

Alternatively, one could also argue that since $\mathbf{b} \in N(A^T)^{\perp}$ and the columns of W form an orthonormal basis for $N(A^T)$

$$W^T\mathbf{b} = \mathbf{0}$$

and hence it follows that

$$QQ^T\mathbf{b} = (I - WW^T)\mathbf{b} = \mathbf{b}$$

(e) If $\mathbf{q}$ is the projection of $\mathbf{c}$ onto $R(A)$ and $\mathbf{r} = \mathbf{c} - \mathbf{q}$, then

$$\mathbf{c} = \mathbf{q} + \mathbf{r}$$

and $\mathbf{r}$ is the projection of $\mathbf{c}$ onto $N(A^T)$.

(f) Since the projection of a vector onto a subspace is unique, $\mathbf{w}$ must equal $\mathbf{r}$.

(g) To compute the projection matrix U, set

$$U = Y * Y'$$

Since $\mathbf{y}$ is already in $R(A^T)$, the projection matrix U should have no effect on $\mathbf{y}$. Thus $U\mathbf{y} = \mathbf{y}$. The vector $\mathbf{s} = \mathbf{b} - \mathbf{y}$ is the projection of $\mathbf{b}$ onto $R(A)^{\perp} = N(A)$. Thus $\mathbf{s} \in N(A)$ and $A\mathbf{s} = \mathbf{0}$.

(h) The vectors $\mathbf{s}$ and $V\mathbf{b}$ should be equal since they are both projections of $\mathbf{b}$ onto $N(A)$.

CHAPTER TEST A

1. The statement is false. The statement is true for nonorthogonal vectors, however, if $\mathbf{x} \perp \mathbf{y}$, then the projection of $\mathbf{x}$ onto $\mathbf{y}$ and the projection of $\mathbf{x}$ onto $\mathbf{x}$ are both equal to $\mathbf{0}$.

2. The statement is false. If $\mathbf{x}$ and $\mathbf{y}$ are unit vectors and θ is the angle between the two vectors, then the condition $|\mathbf{x}^T\mathbf{y}| = 1$ implies that $\cos\theta = \pm 1$. Thus $\mathbf{y} = \mathbf{x}$ or $\mathbf{y} = -\mathbf{x}$. So the vectors $\mathbf{x}$ and $\mathbf{y}$ are linearly dependent.

3. The statement is false. For example, consider the one-dimensional subspaces

$$U = \text{Span}(\mathbf{e}_1), \quad V = \text{Span}(\mathbf{e}_3), \quad W = \text{Span}(\mathbf{e}_1 + \mathbf{e}_2)$$

Since $\mathbf{e}_1 \perp \mathbf{e}_3$ and $\mathbf{e}_3 \perp (\mathbf{e}_1 + \mathbf{e}_2)$, it follows that $U \perp V$ and $V \perp W$. However $\mathbf{e}_1$ is not orthogonal to $\mathbf{e}_1 + \mathbf{e}_2$, so U and W are not orthogonal subspaces.

4. The statement is false. If $\mathbf{y}$ is in the column space of and $A^T\mathbf{y} = \mathbf{0}$, then $\mathbf{y}$ is also in $N(A^T)$. But $R(A) \bigcap N(A^T) = \{\mathbf{0}\}$. So $\mathbf{y}$ must be the zero vector.

5. The statement is true. The matrices A and A^TA have the same rank. (See Exercise 13 of Section 2.) Similarly, A^T and AA^T have the same rank. By Theorem 3.6.6 the matrices A and A^T have the same rank. It follows then that

$$\text{rank}(A^TA) = \text{rank}(A) = \text{rank}(A^T) = \text{rank}(AA^T)$$

6. The statement is false. Although the least squares problem will not have a unique solution the projection of a vector onto any subspace is always unique. See Theorem 5.3.1 or Theorem 5.5.8.

7. The statement is true. If A is $m \times n$ and $N(A) = \{\mathbf{0}\}$, then A has rank n and it follows from Theorem 5.3.2 that the least squares problem will have a unique solution.

8. The statement is true. In general an $n \times n$ matrix Q is orthogonal if and only if $Q^TQ = I$. If Q_1 and Q_2 are both $n \times n$ orthogonal matrices, then

$$(Q_1Q_2)^T(Q_1Q_2) = Q_2^TQ_1^TQ_1Q_2 = Q_2^TIQ_2 = Q_2^TQ_2 = I$$

Therefore Q_1Q_2 is an orthogonal matrix.

9. The statement is true. The matrix U^TU is a $k \times k$ and its (i,j) entry is $\mathbf{u}_i^T\mathbf{u}_j$. Since $\mathbf{u}_1, \mathbf{u}_2, \ldots, \mathbf{u}_k$ are orthonormal vectors, $\mathbf{u}_i^T\mathbf{u}_j = 1$ if $i = j$ and it is equal to 0 otherwise.

10. The statement is false. The statement is only true in the case $k = n$. In the case $k < n$ if we extend the given set of vectors to an orthonormal basis $\{\mathbf{u}_1, \mathbf{u}_2, \ldots, \mathbf{u}_n\}$ for R^n and set

$$V = (\mathbf{u}_{k+1}, \ldots, \mathbf{u}_n), \quad W = (U \; V)$$

then W is an orthogonal matrix and

$$I = WW^T = UU^T + VV^T$$

So UU^T is actually equal to $I - VV^T$. As an example let

$$U = \begin{pmatrix} \frac{1}{3} & \frac{2}{3} \\ \frac{2}{3} & \frac{1}{3} \\ \frac{2}{3} & -\frac{2}{3} \end{pmatrix}$$

The column vectors $\mathbf{u}_1$ and $\mathbf{u}_2$ form an orthonormal set and

$$UU^T = \begin{pmatrix} \frac{1}{3} & \frac{2}{3} \\ \frac{2}{3} & \frac{1}{3} \\ \frac{2}{3} & -\frac{2}{3} \end{pmatrix} \begin{pmatrix} \frac{1}{3} & \frac{2}{3} & \frac{2}{3} \\ \frac{2}{3} & \frac{1}{3} & -\frac{2}{3} \end{pmatrix} = \begin{pmatrix} \frac{5}{9} & \frac{4}{9} & -\frac{2}{9} \\ \frac{3}{9} & \frac{5}{9} & \frac{2}{9} \\ -\frac{2}{9} & \frac{2}{9} & \frac{8}{9} \end{pmatrix}$$

Thus $UU^T \neq I$. Note that if we set

$$\mathbf{u}_3 = \begin{pmatrix} -\frac{2}{3} \\ \frac{2}{3} \\ -\frac{1}{3} \end{pmatrix}$$

then $\{\mathbf{u}_1, \mathbf{u}_2, \mathbf{u}_3\}$ is an orthonormal basis for R^3 and

$$UU^T + \mathbf{u}_3\mathbf{u}_3^T = \begin{pmatrix} \frac{5}{9} & \frac{4}{9} & -\frac{2}{9} \\ \frac{4}{9} & \frac{5}{9} & \frac{2}{9} \\ -\frac{2}{9} & \frac{2}{9} & \frac{8}{9} \end{pmatrix} + \begin{pmatrix} \frac{4}{9} & -\frac{4}{9} & \frac{2}{9} \\ -\frac{4}{9} & \frac{4}{9} & -\frac{2}{9} \\ \frac{2}{9} & -\frac{2}{9} & \frac{1}{9} \end{pmatrix} = I$$

CHAPTER TEST B

1. (a) $\mathbf{p} = \dfrac{\mathbf{x}^T\mathbf{y}}{\mathbf{y}^T\mathbf{y}}\mathbf{y} = \dfrac{3}{9}\mathbf{y} = \left(-\dfrac{2}{3}, \dfrac{1}{3}, \dfrac{2}{3}, 0\right)^T$

(b) $\mathbf{x} - \mathbf{p} = \left(\dfrac{5}{3}, \dfrac{2}{3}, \dfrac{4}{3}, 2\right)^T$

$$(\mathbf{x} - \mathbf{p})^T\mathbf{p} = -\frac{10}{9} + \frac{2}{9} + \frac{8}{9} + 0 = 0$$

(c) $\|\mathbf{x}\|^2 = 1 + 1 + 4 + 4 = 10$

$$\|\mathbf{p}\|^2 + \|\mathbf{x} - \mathbf{p}\|^2 = \left(\frac{4}{9} + \frac{1}{9} + \frac{4}{9} + 0\right) + \left(\frac{25}{9} + \frac{4}{9} + \frac{16}{9} + 4\right) = 1 + 9 = 10$$

2. (a) By the Cauchy-Schwarz inequality

$$|\langle \mathbf{v}_1, \mathbf{v}_2 \rangle| \leq \|\mathbf{v}_1\|\|\mathbf{v}_2\|$$

(b) If

$$|\langle \mathbf{v}_1, \mathbf{v}_2 \rangle| = \|\mathbf{v}_1\|\|\mathbf{v}_2\|$$

then equality holds in the Cauchy-Schwarz inequality and this can only happen if the two vectors are linearly dependent.

3.

$$\begin{aligned}\|\mathbf{v}_1+\mathbf{v}_2\|^2 &= \langle \mathbf{v}_1+\mathbf{v}_2, \mathbf{v}_1+\mathbf{v}_2\rangle \\ &= \langle \mathbf{v}_1,\mathbf{v}_1\rangle + 2\langle \mathbf{v}_1,\mathbf{v}_2\rangle + \langle \mathbf{v}_2,\mathbf{v}_2\rangle \\ &\le \|\mathbf{v}_1\|^2 + 2\|\mathbf{v}_1\|\|\mathbf{v}_2\| + \|\mathbf{v}_2\|^2 \quad \text{(Cauchy – Schwarz)} \\ &= (\|\mathbf{v}_1\| + \|\mathbf{v}_2\|)^2\end{aligned}$$

4. (a) IfA has rank 4 then A^T must also have rank 4. The matrix A^T has 7 columns, so by the Rank-Nullity theorem its rank and nullity must add up to 7. Since the rank is 4, the nullity must be 3 and hence $\dim N(A^T) = 3$. The orthogonal complement of $N(A^T)$ is $R(A)$.

(b) If $\mathbf{x}$ is in $R(A)$ and $A^T\mathbf{x} = \mathbf{0}$ then $\mathbf{x}$ is also in $N(A^T)$. Since $R(A)$ and $N(A^T)$ are orthogonal subspaces their intersection is $\{\mathbf{0}\}$. Therefore $\mathbf{x} = \mathbf{0}$ and $\|\mathbf{x}\| = 0$.

(c) $\dim N(A^TA) = \dim N(A) = 1$ by the Rank-Nullity Theorem. Therefore the normal equations will involve 1 free variables and hence the least squares problem will have infinitely many solutions.

5. If θ_1 is the angle between $\mathbf{x}$ and $\mathbf{y}$ and θ_2 is the angle between $Q\mathbf{x}$ and $Q\mathbf{y}$ then

$$\cos\theta_2 = \frac{(Q\mathbf{x})^TQ\mathbf{y}}{\|Q\mathbf{x}\|\,\|Q\mathbf{y}\|} = \frac{\mathbf{x}^TQ^TQ\mathbf{y}}{\|\mathbf{x}\|\,\|\mathbf{y}\|} = \frac{\mathbf{x}^T\mathbf{y}}{\|\mathbf{x}\|\,\|\mathbf{y}\|} = \cos\theta_1$$

The angles θ_1 and θ_2 must both be in the interval $[0, \pi]$. Since their cosines are equal, the angles must be equal.

6. (a) If we let $X = (\mathbf{x}_1, \mathbf{x}_2)$ then $S = R(X)$ and hence

$$S^\perp = R(X)^\perp = N(X^T)$$

To find a basis for $S^\perp$ we solve $X^T\mathbf{x} = \mathbf{0}$. The matrix

$$X^T = \begin{pmatrix} 1 & 0 & 2 \\ 0 & 1 & -2 \end{pmatrix}$$

is already in reduced row echelon form with one free variable $\mathbf{x}_3$. If we set $x_3 = a$, then $x_1 = -2a$ and $x_2 = 2a$. Thus $S^\perp$ consists of all vectors of the form $(-2a, 2a, a)^T$ and $\{(-2, 2, 1)^T\}$ is a basis for $S^\perp$.

(b) S is the span of two linearly independent vectors and hence S can be represented geometrically by a plane through the origin in 3-space. $S^\perp$ corresponds to the line through the original that is normal to the plane representing S.

(c) To find the projection matrix we must find an orthonormal basis for $S^\perp$. Since $\dim S^\perp = 1$ we need only normalize our single basis vector to obtain an orthonormal basis. If we set $\mathbf{u} = \frac{1}{3}(-2, 2, 1)^T$ then the projection matrix is

$$P = \mathbf{u}\mathbf{u}^T = \frac{1}{9}\begin{pmatrix} -2 \\ 2 \\ 1 \end{pmatrix}\begin{pmatrix} -2 & 2 & 1 \end{pmatrix} = \begin{pmatrix} \frac{4}{9} & -\frac{4}{9} & -\frac{2}{9} \\ -\frac{4}{9} & \frac{4}{9} & \frac{2}{9} \\ -\frac{2}{9} & \frac{2}{9} & \frac{1}{9} \end{pmatrix}$$

7. To find the best least squares fit we must find a least squares solution to the system

$$\begin{aligned} c_1 - c_2 &= 1 \\ c_1 + c_2 &= 3 \\ c_1 + 2c_2 &= 3 \end{aligned}$$

If A is the coefficient matrix for this system and $\mathbf{b}$ is the right hand side, then the solution $\mathbf{c}$ to the least squares problem is the solution to the normal equations $A^TA\mathbf{c} = A^T\mathbf{b}$.

$$A^TA = \begin{pmatrix} 1 & 1 & 1 \\ -1 & 1 & 2 \end{pmatrix} \begin{pmatrix} 1 & -1 \\ 1 & 1 \\ 1 & 2 \end{pmatrix} = \begin{pmatrix} 3 & 2 \\ 2 & 6 \end{pmatrix}$$

$$A^T\mathbf{b} = \begin{pmatrix} 1 & 1 & 1 \\ -1 & 1 & 2 \end{pmatrix} \begin{pmatrix} 1 \\ 3 \\ 3 \end{pmatrix} = \begin{pmatrix} 7 \\ 8 \end{pmatrix}$$

The augmented matrix for the normal equations is

$$\left(\begin{array}{cc|c} 3 & 2 & 7 \\ 2 & 6 & 8 \end{array}\right)$$

The solution to this system is $\mathbf{c} = (\frac{13}{7}, \frac{5}{7})^T$ and hence the best linear fit is $f(x) = \frac{13}{7} + \frac{5}{7}x$.

8. **(a)** It follows from Theorem 5.5.3 that

$$\langle \mathbf{x}, \mathbf{y} \rangle = 2 \cdot 3 + (-2) \cdot 1 + 1 \cdot (-4) = 0$$

(so $\mathbf{x}$ and $\mathbf{y}$ are orthogonal).

(b) By Parseval's formula

$$\|\mathbf{x}\|^2 = 2^2 + (-2)^2 + 1^2 = 9$$

and therefore $\|\mathbf{x}\| = 3$.

9. **(a)** If $\mathbf{x}$ is any vector in $N(A^T)$ then $\mathbf{x}$ is in $R(A)^\perp$ and hence the projection of $\mathbf{x}$ onto $R(A)$ will be $\mathbf{0}$, i.e., $P\mathbf{x} = \mathbf{0}$. The column vectors of Q are all in $N(A^T)$ since Q projects vectors onto $N(A^T)$ and $\mathbf{q}_j = Q\mathbf{e}_j$ for $1 \le j \le 7$. It follows then that

$$PQ = (P\mathbf{q}_1, P\mathbf{q}_2, P\mathbf{q}_3, P\mathbf{q}_4, P\mathbf{q}_5, P\mathbf{q}_6, P\mathbf{q}_7) = (\mathbf{0}, \mathbf{0}, \mathbf{0}, \mathbf{0}, \mathbf{0}, \mathbf{0}, \mathbf{0}) = O$$

(b) Let $\{\mathbf{u}_1, \mathbf{u}_2, \mathbf{u}_3, \mathbf{u}_4\}$ is an orthonormal basis for $R(A)$ and let $\{\mathbf{u}_5, \mathbf{u}_6, \mathbf{u}_7\}$ be an orthonormal basis for $N(A^T)$. If we set $U_1 = (\mathbf{u}_1, \mathbf{u}_2, \mathbf{u}_3, \mathbf{u}_4)$ and $U_2 = (\mathbf{u}_5, \mathbf{u}_6, \mathbf{u}_7)$ then $P = U_1U_1^T$ and $Q = U_2U_2^T$. The matrix $U = (U_1, U_2)$ is orthogonal and hence $U^{-1} = U^T$. It follows then that

$$I = UU^T = \begin{pmatrix} U_1 & U_2 \end{pmatrix} \begin{pmatrix} U_1^T \\ U_2^T \end{pmatrix} = U_1U_1^T + U_2U_2^T = P + Q$$

10. **(a)** $r_{13} = \mathbf{q}_1^T\mathbf{a}_3 = -1$, $r_{23} = \mathbf{q}_2^T\mathbf{a}_3 = 3$, $\mathbf{p}_2 = -\mathbf{q}_1 + 3\mathbf{q}_2 = (-2, 1, -2, 1)^T$

$\mathbf{a}_3 - \mathbf{p}_2 = (-3, -3, 3, 3)^T$, $r_{33} = \|\mathbf{a}_3 - \mathbf{p}_2\| = 6$

$\mathbf{q}_3 = \frac{1}{6}(-3, -3, 3, 3)^T = (-\frac{1}{2}, -\frac{1}{2}, \frac{1}{2}, \frac{1}{2})^T$

(b)

$$\mathbf{c} = Q^T\mathbf{b} = \begin{pmatrix} \frac{1}{2} & \frac{1}{2} & \frac{1}{2} & \frac{1}{2} \\ -\frac{1}{2} & \frac{1}{2} & -\frac{1}{2} & \frac{1}{2} \\ -\frac{1}{2} & -\frac{1}{2} & \frac{1}{2} & \frac{1}{2} \end{pmatrix} \begin{pmatrix} -6 \\ 1 \\ 1 \\ 6 \end{pmatrix} = \begin{pmatrix} 1 \\ 6 \\ 6 \end{pmatrix}$$

To solve the least squares problem we must solve the upper triangular system $R\mathbf{x} = \mathbf{c}$. The augmented matrix for this system is

$$\left(\begin{array}{ccc|c} 2 & -2 & -1 & 1 \\ 0 & 4 & 3 & 6 \\ 0 & 0 & 6 & 6 \end{array}\right)$$

and the solution $\mathbf{x} = (\frac{7}{4}, \frac{3}{4}, 1)^T$ is easily obtained using back substitution.

11. **(a)** $\langle \cos x, \sin x \rangle = \frac{1}{\pi} \int_{-\pi}^{\pi} \cos x \sin x \, dx = 0$

(b) Since $\cos x$ and $\sin x$ are orthogonal we have by the Pythagorean Law that

$$\begin{aligned} \|\cos x + \sin x\|^2 &= \|\cos x\|^2 + \|\sin x\|^2 \\ &= \frac{1}{\pi}\int_{-\pi}^{\pi} \cos^2 x \, dx + \frac{1}{\pi}\int_{-\pi}^{\pi} \sin^2 x \, dx \\ &= \frac{1}{\pi}\int_{-\pi}^{\pi} 1 \, dx = 2 \end{aligned}$$

Therefore $\|\cos x + \sin x\| = \sqrt{2}$.

12. **(a)** $\langle u_1(x), u_2(x) \rangle = \int_{-1}^{1} \frac{1}{\sqrt{2}} \frac{\sqrt{6}}{2} x \, dx = 0$

$\langle u_1(x), u_1(x) \rangle = \int_{-1}^{1} \frac{1}{2} \, dx = 1$

$\langle u_2(x), u_2(x) \rangle = \int_{-1}^{1} \frac{3}{2} x^2 \, dx = 1$

(b) Let

$$\begin{aligned} c_1 &= \langle h(x), u_1(x) \rangle = \frac{1}{\sqrt{2}} \int_{-1}^{1} (x^{1/3} + x^{2/3}) \, dx = \frac{6}{5\sqrt{2}} \\ c_2 &= \langle h(x), u_2(x) \rangle = \frac{\sqrt{6}}{2} \int_{-1}^{1} (x^{1/3} + x^{2/3}) x \, dx = \frac{3\sqrt{6}}{7} \end{aligned}$$

The best linear approximation to $h(x)$ is

$$f(x) = c_1 u_1(x) + c_2 u_2(x) = \frac{3}{5} + \frac{9}{7}x$$

Chapter 6

SECTION 1

2. If A is triangular then $A - a_{ii}I$ will be a triangular matrix with a zero entry in the (i, i) position. Since the determinant of a triangular matrix is the product of its diagonal elements it follows that

$$\det(A - a_{ii}I) = 0$$

Thus the eigenvalues of A are $a_{11}, a_{22}, \ldots, a_{nn}$.

3. A is singular if and only if $\det(A) = 0$. The scalar 0 is an eigenvalue if and only if

$$\det(A - 0I) = \det(A) = 0$$

Thus A is singular if and only if one of its eigenvalues is 0.

4. If A is a nonsingular matrix and λ is an eigenvalue of A, then there exists a nonzero vector $\mathbf{x}$ such that

$$\begin{aligned} A\mathbf{x} &= \lambda\mathbf{x} \\ A^{-1}A\mathbf{x} &= \lambda A^{-1}\mathbf{x} \end{aligned}$$

It follows from Exercise 3 that $\lambda \neq 0$. Therefore

$$A^{-1}\mathbf{x} = \frac{1}{\lambda}\mathbf{x} \quad (\mathbf{x} \neq \mathbf{0})$$

and hence $1/\lambda$ is an eigenvalue of A^{-1}.

5. The proof is by induction. In the case where $m = 1$, $\lambda^1 = \lambda$ is an eigenvalue of A with eigenvector $\mathbf{x}$. Suppose λ^k is an eigenvalue of A^k and $\mathbf{x}$ is an eigenvector belonging to λ^k.

$$A^{k+1}\mathbf{x} = A(A^k\mathbf{x}) = A(\lambda^k\mathbf{x}) = \lambda^k A\mathbf{x} = \lambda^{k+1}\mathbf{x}$$

Thus λ^{k+1} is an eigenvalue of A^{k+1} and $\mathbf{x}$ is an eigenvector belonging to λ^{k+1}. It follows by induction that if λ an eigenvalue of A then λ^m is an eigenvalue of A^m, for $m = 1, 2, \ldots$.

6. If A is idempotent and λ is an eigenvalue of A with eigenvector $\mathbf{x}$, then

$$\begin{aligned} A\mathbf{x} &= \lambda\mathbf{x} \\ A^2\mathbf{x} &= \lambda A\mathbf{x} = \lambda^2\mathbf{x} \end{aligned}$$

and

$$A^2\mathbf{x} = A\mathbf{x} = \lambda\mathbf{x}$$

Therefore

$$(\lambda^2 - \lambda)\mathbf{x} = \mathbf{0}$$

Since $\mathbf{x} \neq \mathbf{0}$ it follows that

$$\begin{aligned} &\lambda^2 - \lambda = 0 \\ &\lambda = 0 \quad \text{or} \quad \lambda = 1 \end{aligned}$$

7. If λ is an eigenvalue of A, then λ^k is an eigenvalue of A^k (Exercise 5). If $A^k = O$, then all of its eigenvalues are 0. Thus $\lambda^k = 0$ and hence $\lambda = 0$.

9. $\det(A - \lambda I) = \det((A - \lambda I)^T) = \det(A^T - \lambda I)$. Thus A and A^T have the same characteristic polynomials and consequently must have the same eigenvalues. The eigenspaces however will not be the same. For example

$$A = \begin{pmatrix} 1 & 1 \\ 0 & 1 \end{pmatrix} \quad \text{and} \quad A^T = \begin{pmatrix} 1 & 0 \\ 1 & 1 \end{pmatrix}$$

both have eigenvalues

$$\lambda_1 = \lambda_2 = 1$$

The eigenspace of A corresponding to $\lambda = 1$ is spanned by $(1,\ 0)^T$ while the eigenspace of A^T is spanned by $(0,\ 1)^T$. Exercise 27 shows how the eigenvectors of A and A^T are related.

10. $\det(A - \lambda I) = \lambda^2 - (2\cos\theta)\lambda + 1$. The discriminant will be negative unless θ is a multiple of π. The matrix A has the effect of rotating a real vector $\mathbf{x}$ about the origin by an angle of θ. Thus $A\mathbf{x}$ will be a scalar multiple of $\mathbf{x}$ if and only if θ is a multiple of π.

12. Since $\text{tr}(A)$ equals the sum of the eigenvalues the result follows by solving

$$\sum_{i=1}^{n} \lambda_i = \sum_{i=1}^{n} a_{ii}$$

for λ_j.

13. $\begin{vmatrix} a_{11}-\lambda & a_{12} \\ a_{21} & a_{22}-\lambda \end{vmatrix} = \lambda^2 - (a_{11}+a_{22})\lambda + (a_{11}a_{22} - a_{21}a_{12})$

$$= \lambda^2 - (\text{tr}\, A)\lambda + \det(A)$$

14. If $\mathbf{x}$ is an eigenvector of A belonging to λ, then any nonzero multiple of $\mathbf{x}$ is also an eigenvector of A belonging to λ. By Exercise 5 we know that $A^m\mathbf{x} = \lambda^m\mathbf{x}$, so $A^m\mathbf{x}$ must be an eigenvector of A belonging to λ. Alternatively we could have proved the result by noting that

$$A^m\mathbf{x} = \lambda^m\mathbf{x} \neq \mathbf{0}$$

and

$$A(A^m\mathbf{x}) = A^{m+1}\mathbf{x} = A^m(A\mathbf{x}) = A^m(\lambda\mathbf{x}) = \lambda(A^m\mathbf{x})$$

15. If $A - \lambda_0 I$ has rank k then $N(A - \lambda_0 I)$ will have dimension $n - k$.

16. The subspace spanned by $\mathbf{x}$ and $A\mathbf{x}$ will have dimension 1 if and only if $\mathbf{x}$ and $A\mathbf{x}$ are linearly dependent and $\mathbf{x} \neq \mathbf{0}$. If $\mathbf{x} \neq \mathbf{0}$ then the vectors $\mathbf{x}$ and $A\mathbf{x}$ will be linearly dependent if and only if $A\mathbf{x} = \lambda\mathbf{x}$ for some scalar λ.

17. (a) If $\alpha = a + bi$ and $\beta = c + di$, then

$$\overline{\alpha + \beta} = \overline{(a+c) + (b+d)i} = (a+c) - (b+d)i$$

and

$$\overline{\alpha} + \overline{\beta} = (a - bi) + (c - di) = (a+c) - (b+d)i$$

Therefore $\overline{\alpha+\beta} = \overline{\alpha} + \overline{\beta}$.

Next we show that the conjugate of the product of two numbers is the product of the conjugates.

$$\overline{\alpha\beta} = \overline{(ac - bd) + (ad + bc)i} = (ac - bd) - (ad + bc)i$$

$$\overline{\alpha}\,\overline{\beta} = (a - bi)(c - di) = (ac - bd) - (ad + bc)i$$

Therefore $\overline{\alpha\beta} = \overline{\alpha}\,\overline{\beta}$.

(b) If $A \in R^{m\times n}$ and $B \in R^{n\times r}$, then the (i,j) entry of $\overline{AB}$ is given by

$$\overline{a_{i1}b_{1j} + a_{i2}b_{2j} + \cdots + a_{in}b_{nj}} = \overline{a_{i1}}\,\overline{b_{1j}} + \overline{a_{i2}}\,\overline{b_{2j}} + \cdots + \overline{a_{in}}\,\overline{b_{nj}}$$

The expression on the right is the (i,j) entry of $\overline{A}\,\overline{B}$. Therefore

$$\overline{AB} = \overline{A}\,\overline{B}$$

18. **(a)** If λ is an eigenvalue of an orthogonal matrix Q and $\mathbf{x}$ is a unit eigenvector belonging to λ then

$$|\lambda| = |\lambda|\,\|\mathbf{x}\| = \|\lambda\mathbf{x}\| = \|Q\mathbf{x}\| = \|\mathbf{x}\| = 1$$

(b) Since the eigenvalues of Q all have absolute value equal to 1, it follows that

$$|\det(Q)| = |\lambda_1\lambda_2\cdots\lambda_n| = 1$$

19. If Q is an orthogonal matrix with eigenvalue $\lambda = 1$ and $\mathbf{x}$ is an eigenvector belonging to $\lambda = 1$, then $Q\mathbf{x} = \mathbf{x}$ and since $Q^T = Q^{-1}$ we have

$$Q^T\mathbf{x} = Q^TQ\mathbf{x} = I\mathbf{x} = \mathbf{x}$$

Therefore $\mathbf{x}$ is an eigenvector of Q^T belonging to the eigenvector $\lambda = 1$.

20. **(a)** Each eigenvalue has absolute value 1 and the product of the eigenvalues is equal to 1. So if the eigenvalues are real and are ordered so that $\lambda_1 \geq \lambda_2 \geq \lambda_3$, then the only possible triples of eigenvalues are: $(1, 1, 1)$ and $(1, -1, -1)$.

(b) The complex eigenvalues must be of the form $\lambda_2 = \cos\theta + i\sin\theta$ and $\lambda_3 = \cos\theta - i\sin\theta$. It follows then that

$$\lambda_1\lambda_2\lambda_3 = \lambda_1(\cos\theta + i\sin\theta)(\cos\theta - i\sin\theta) = \lambda_1(\cos^2\theta + \sin^2\theta) = \lambda_1$$

Therefore

$$\lambda_1 = \lambda_1\lambda_2\lambda_3 = \det(A) = 1$$

(c) If the eigenvalues of Q are all real then by part (a) at least one of the eigenvalues must equal 1. If the eigenvalues are not all real then Q must have one pair of complex conjugate eigenvalues and one real eigenvalue. By part (b) the real eigenvalue must be equal to 1. Therefore if Q is a 3×3 orthogonal matrix with $\det(Q) = 1$, then $\lambda = 1$ must be an eigenvalue.

21. If $\mathbf{x} = c_1\mathbf{x}_1 + c_2\mathbf{x}_2 + \cdots + c_r\mathbf{x}_r$ is an element of S, then

$$A\mathbf{x} = (c_1\lambda_1)\mathbf{x}_1 + (c_2\lambda_2)\mathbf{x}_2 + \cdots + (c_r\lambda_r)\mathbf{x}_r$$

Thus $A\mathbf{x}$ is also an element of S.

22. Since $\mathbf{x} \neq \mathbf{0}$ and S is nonsingular it follows that $S\mathbf{x} \neq \mathbf{0}$. If $B = S^{-1}AS$, then $AS = SB$ and it follows that

$$A(S\mathbf{x}) = (AS)\mathbf{x} = SB\mathbf{x} = S(\lambda\mathbf{x}) = \lambda(S\mathbf{x})$$

Therefore $S\mathbf{x}$ is an eigenvector of A belonging to λ.

23. If $\mathbf{x}$ is an eigenvector of A belonging to the eigenvalue λ and $\mathbf{x}$ is also an eigenvector of B corresponding to the eigenvalue μ, then

$$(\alpha A + \beta B)\mathbf{x} = \alpha A\mathbf{x} + \beta B\mathbf{x} = \alpha\lambda\mathbf{x} + \beta\mu\mathbf{x} = (\alpha\lambda + \beta\mu)\mathbf{x}$$

Therefore $\mathbf{x}$ is an eigenvector of $\alpha A + \beta B$ belonging to $\alpha\lambda + \beta\mu$.

24. If $\lambda \neq 0$ and $\mathbf{x}$ is an eigenvector belonging to λ, then

$$\begin{aligned} A\mathbf{x} &= \lambda\mathbf{x} \\ \mathbf{x} &= \frac{1}{\lambda}A\mathbf{x} \end{aligned}$$

Since $A\mathbf{x}$ is in $R(A)$ it follows that $\frac{1}{\lambda}A\mathbf{x}$ is in $R(A)$.

25. If

$$A = \lambda_1\mathbf{u}_1\mathbf{u}_1^T + \lambda_2\mathbf{u}_2\mathbf{u}_2^T + \cdots + \lambda_n\mathbf{u}_n\mathbf{u}_n^T$$

then for $i = 1, \ldots, n$

$$A\mathbf{u}_i = \lambda_1\mathbf{u}_1\mathbf{u}_1^T\mathbf{u}_i + \lambda_2\mathbf{u}_2\mathbf{u}_2^T\mathbf{u}_i + \cdots + \lambda_n\mathbf{u}_n\mathbf{u}_n^T\mathbf{u}_i$$

Since $\mathbf{u}_j^T\mathbf{u}_i = 0$ unless $j = i$, it follows that

$$A\mathbf{u}_i = \lambda_i\mathbf{u}_i\mathbf{u}_i^T\mathbf{u}_i = \lambda_i\mathbf{u}_i$$

and hence λ_i is an eigenvalue of A with eigenvector $\mathbf{u}_i$. The matrix A is symmetric since each $c_i\mathbf{u}_i\mathbf{u}_i^T$ is symmetric and any sum of symmetric matrices is symmetric.

26. If the columns of A each add up to a fixed constant δ then the row vectors of $A-\delta I$ all add up to $(0,0,\ldots,0)$. Thus the row vectors of $A-\delta I$ are linearly dependent and hence $A-\delta I$ is singular. Therefore δ is an eigenvalue of A.

27. Since $\mathbf{y}$ is an eigenvector of A^T belonging to λ_2 it follows that

$$\mathbf{x}^T A^T \mathbf{y} = \lambda_2 \mathbf{x}^T \mathbf{y}$$

The expression $\mathbf{x}^T A^T \mathbf{y}$ can also be written in the form $(A\mathbf{x})^T\mathbf{y}$. Since $\mathbf{x}$ is an eigenvector of A belonging to λ_1, it follows that

$$\mathbf{x}^T A^T \mathbf{y} = (A\mathbf{x})^T\mathbf{y} = \lambda_1 \mathbf{x}^T \mathbf{y}$$

Therefore

$$(\lambda_1 - \lambda_2)\mathbf{x}^T\mathbf{y} = 0$$

and since $\lambda_1 \neq \lambda_2$, the vectors $\mathbf{x}$ and $\mathbf{y}$ must be orthogonal.

28. (a) If λ is a nonzero eigenvalue of AB with eigenvector $\mathbf{x}$, then let $\mathbf{y} = B\mathbf{x}$. Since

$$A\mathbf{y} = AB\mathbf{x} = \lambda\mathbf{x} \neq \mathbf{0}$$

it follows that $\mathbf{y} \neq \mathbf{0}$ and

$$BA\mathbf{y} = BA(B\mathbf{x}) = B(AB\mathbf{x}) = B\lambda\mathbf{x} = \lambda\mathbf{y}$$

Thus λ is also an eigenvalue of BA with eigenvector $\mathbf{y}$.

(b) If $\lambda = 0$ is an eigenvalue of AB, then AB must be singular. Since

$$\det(BA) = \det(B)\det(A) = \det(A)\det(B) = \det(AB) = 0$$

it follows that BA is also singular. Therefore $\lambda = 0$ is an eigenvalue of BA.

29. If $AB - BA = I$, then $BA = AB - I$. If the eigenvalues of AB are $\lambda_1, \lambda_2, \ldots, \lambda_n$, then it follows from Exercise 8 that the eigenvalues of BA are $\lambda_1 - 1$, $\lambda_2 - 1, \ldots, \lambda_n - 1$. This contradicts the result proved in Exercise 28 that AB and BA have the same eigenvalues.

30. (a) If λ_i is a root of $p(\lambda)$, then

$$\lambda_i^n = a_{n-1}\lambda_i^{n-1} + \cdots + a_1\lambda_i + a_0$$

Thus if $\mathbf{x} = (\lambda_i^{n-1}, \lambda_i^{n-2}, \ldots, \lambda_i, 1)^T$, then

$$C\mathbf{x} = (\lambda_i^n, \lambda_i^{n-1}, \ldots, \lambda_i^2, \lambda_i)^T = \lambda_i\mathbf{x}$$

and hence λ_i is an eigenvalue of C with eigenvector $\mathbf{x}$.

(b) If $\lambda_1, \ldots, \lambda_n$ are the roots of $p(\lambda)$, then

$$p(\lambda) = (-1)^n(\lambda - \lambda_1)\cdots(\lambda - \lambda_n)$$

If $\lambda_1, \ldots, \lambda_n$ are all distinct then by part (a) they are the eigenvalues of C. Since the characteristic polynomial of C has lead coefficient $(-1)^n$ and roots $\lambda_1, \ldots, \lambda_n$, it must equal $p(\lambda)$.

31. Let

$$D_m(\lambda) = \begin{pmatrix} a_m & a_{m-1} & \cdots & a_1 & a_0 \\ 1 & -\lambda & \cdots & 0 & 0 \\ \vdots & & & & \\ 0 & 0 & \cdots & 1 & -\lambda \end{pmatrix}$$

It can be proved by induction on m that

$$\det(D_m(\lambda)) = (-1)^m(a_m\lambda^m + a_{m-1}\lambda^{m-1} + \cdots + a_1\lambda + a_0)$$

If $\det(C - \lambda I)$ is expanded by cofactors along the first column one obtains

$$\begin{aligned} \det(C - \lambda I) &= (a_{n-1} - \lambda)(-\lambda)^{n-1} - \det(D_{n-2}) \\ &= (-1)^n(\lambda^n - a_{n-1}\lambda^{n-1}) - (-1)^{n-2}(a_{n-2}\lambda^{n-2} + \cdots + a_1\lambda + a_0) \\ &= (-1)^n[(\lambda^n - a_{n-1}\lambda^{n-1}) - (a_{n-2}\lambda^{n-2} + \cdots + a_1\lambda + a_0)] \\ &= (-1)^n[\lambda^n - a_{n-1}\lambda^{n-1} - a_{n-2}\lambda^{n-2} - \cdots - a_1\lambda - a_0] \\ &= p(\lambda) \end{aligned}$$

SECTION 2

3. (a) If

$$\mathbf{Y}(t) = c_1e^{\lambda_1 t}\mathbf{x}_1 + c_2e^{\lambda_2 t}\mathbf{x}_2 + \cdots + c_ne^{\lambda_n t}\mathbf{x}_n$$

then

$$\mathbf{Y}_0 = \mathbf{Y}(0) = c_1\mathbf{x}_2 + c_2\mathbf{x}_2 + \cdots + c_n\mathbf{x}_n$$

(b) It follows from part (a) that

$$\mathbf{Y}_0 = X\mathbf{c}$$

If $\mathbf{x}_1, \ldots, \mathbf{x}_n$ are linearly independent then X is nonsingular and we can solve for $\mathbf{c}$

$$\mathbf{c} = X^{-1}\mathbf{Y}_0$$

7. It follows from the initial condition that

$$\begin{aligned} x_1'(0) &= a_1\sigma = 2 \\ x_2'(0) &= a_2\sigma = 2 \end{aligned}$$

and hence

$$a_1 = a_2 = 2/\sigma$$

Substituting for x_1 and x_2 in the system

$$\begin{aligned} x_1'' &= -2x_1 + x_2 \\ x_2'' &= x_1 - 2x_2 \end{aligned}$$

yields

$$\begin{aligned} -a_1\sigma^2 \sin\sigma t &= -2a_1 \sin\sigma t + a_2 \sin\sigma t \\ -a_2\sigma^2 \sin\sigma t &= a_1 \sin\sigma t - 2a_2 \sin\sigma t \end{aligned}$$

Replacing a_1 and a_2 by $2/\sigma$ we get

$$\sigma^2 = 1$$

Using either $\sigma = -1$, $a_1 = a_2 = -2$ or $\sigma = 1$, $a_1 = a_2 = 2$ we obtain the solution

$$\begin{aligned} x_1(t) &= 2\sin t \\ x_2(t) &= 2\sin t \end{aligned}$$

9. $m_1 y_1'' = k_1 y_1 - k_2(y_2 - y_1) - m_1 g$
$m_2 y_2'' = k_2(y_2 - y_1) - m_2 g$

11. If

$$y^{(n)} = a_0 y + a_1 y' + \cdots + a_{n-1} y^{(n-1)}$$

and we set

$$y_1 = y,\ y_2 = y_1' = y'',\ y_3 = y_2' = y''', \ldots, y_n = y_{n-1}' = y^n$$

then the nth order equation can be written as a system of first order equations of the form $\mathbf{Y}' = A\mathbf{Y}$ where

$$A = \begin{pmatrix} 0 & y_2 & 0 & \cdots & 0 \\ 0 & 0 & y_3 & \cdots & 0 \\ \vdots & & & & \\ 0 & 0 & 0 & \cdots & y_n \\ a_0 & a_1 & a_2 & \cdots & a_{n-1} \end{pmatrix}$$

SECTION 3

1. The factorization XDX^{-1} is not unique. However the diagonal elements of D must be eigenvalues of A and if λ_i is the ith diagonal element of D, then $\mathbf{x}_i$ must be an eigenvector belonging to λ_i

(a) $\det(A - \lambda I) = \lambda^2 - 1$ and hence the eigenvalues are $\lambda_1 = 1$ and $\lambda_2 = -1$. $\mathbf{x}_1 = (1,\ 1)^T$ and $\mathbf{x}_2 = (-1,\ 1)^T$ are eigenvectors belonging to λ_1 and λ_2, respectively. Setting

$$X = \begin{pmatrix} 1 & -1 \\ 1 & 1 \end{pmatrix} \quad \text{and} \quad D = \begin{pmatrix} 1 & 0 \\ 0 & -1 \end{pmatrix}$$

we have

$$A = XDX^{-1} = \begin{pmatrix} 1 & -1 \\ 1 & 1 \end{pmatrix} \begin{pmatrix} 1 & 0 \\ 0 & -1 \end{pmatrix} \begin{pmatrix} 1/2 & 1/2 \\ -1/2 & 1/2 \end{pmatrix}$$

(b) The eigenvalues are $\lambda_1 = 2$, $\lambda_2 = 1$. If we take $\mathbf{x}_1 = (-2,\ 1)^T$ and $\mathbf{x}_2 = (-3,\ 2)^T$, then

$$A = XDX^{-1} = \begin{pmatrix} -2 & -3 \\ 1 & 2 \end{pmatrix} \begin{pmatrix} 2 & 0 \\ 0 & 1 \end{pmatrix} \begin{pmatrix} -2 & -3 \\ 1 & 2 \end{pmatrix}$$

(c) $\lambda_1 = 0$, $\lambda_2 = -2$. If we take $\mathbf{x}_1 = (4,\ 1)^T$ and $\mathbf{x}_2 = (2,\ 1)^T$, then

$$A = XDX^{-1} = \begin{pmatrix} 4 & 2 \\ 1 & 1 \end{pmatrix} \begin{pmatrix} 0 & 0 \\ 0 & -2 \end{pmatrix} \begin{pmatrix} 1/2 & -1 \\ -1/2 & 2 \end{pmatrix}$$

(d) The eigenvalues are the diagonal entries of A. The eigenvectors corresponding to $\lambda_1 = 2$ are all multiples of $(1,\ 0,\ 0)^T$. The eigenvectors belonging to $\lambda_2 = 1$ are all multiples of $(2,\ -1,\ 0)$ and the eigenvectors corresponding to $\lambda_3 = -1$ are multiples $(1,\ -3,\ 3)^T$.

$$A = XDX^{-1} = \begin{pmatrix} 1 & 2 & 1 \\ 0 & -1 & -3 \\ 0 & 0 & 3 \end{pmatrix} \begin{pmatrix} 2 & 0 & 0 \\ 0 & 1 & 0 \\ 0 & 0 & -1 \end{pmatrix} \begin{pmatrix} 1 & 2 & \frac{5}{3} \\ 0 & -1 & -1 \\ 0 & 0 & \frac{1}{3} \end{pmatrix}$$

(e) $\lambda_1 = 1$, $\lambda_2 = 2$, $\lambda_3 = -2$
$\mathbf{x}_1 = (3,\ 1,\ 2)^T$, $\mathbf{x}_2 = (0,\ 3,\ 1)^T$, $\mathbf{x}_3 = (0,\ -1,\ 1)^T$

$$A = XDX^{-1} = \begin{pmatrix} 3 & 0 & 0 \\ 1 & 3 & -1 \\ 2 & 1 & 1 \end{pmatrix} \begin{pmatrix} 1 & 0 & 0 \\ 0 & 2 & 0 \\ 0 & 0 & -2 \end{pmatrix} \begin{pmatrix} \frac{1}{3} & 0 & 0 \\ -\frac{1}{4} & \frac{1}{4} & \frac{1}{4} \\ -\frac{5}{12} & -\frac{1}{4} & \frac{3}{4} \end{pmatrix}$$

(f) $\lambda_1 = 2$, $\lambda_2 = \lambda_3 = 0$, $\mathbf{x}_1 = (1,\ 2,\ 3)^T$, $\mathbf{x}_2 = (1,\ 0,\ 1)^T$, $\mathbf{x}_3 = (-2,\ 1,\ 0)^T$

$$A = XDX^{-1} = \begin{pmatrix} 1 & 1 & -2 \\ 2 & 0 & 1 \\ 3 & 1 & 0 \end{pmatrix} \begin{pmatrix} 2 & 0 & 0 \\ 0 & 0 & 0 \\ 0 & 0 & 0 \end{pmatrix} \begin{pmatrix} \frac{1}{2} & 1 & -\frac{1}{2} \\ -\frac{3}{2} & -3 & \frac{5}{2} \\ -1 & -1 & 1 \end{pmatrix}$$

2. If $A = XDX^{-1}$, then $A^6 = XD^6X^{-1}$.

(a) $D^6 = \begin{pmatrix} 1 & 0 \\ 0 & -1 \end{pmatrix}^6 = I$

$A^6 = XD^6X^{-1} = XX^{-1} = I$

(b) $A^6 = \begin{pmatrix} -2 & -3 \\ 1 & 2 \end{pmatrix} \begin{pmatrix} 2 & 0 \\ 0 & 1 \end{pmatrix}^6 \begin{pmatrix} -2 & -3 \\ 1 & 2 \end{pmatrix} = \begin{pmatrix} 253 & 378 \\ -126 & -190 \end{pmatrix}$

(c) $A^6 = \begin{pmatrix} 4 & 2 \\ 1 & 1 \end{pmatrix} \begin{pmatrix} 0 & 0 \\ 0 & -2 \end{pmatrix}^6 \begin{pmatrix} 1/2 & -1 \\ -1/2 & 2 \end{pmatrix} = \begin{pmatrix} -64 & 256 \\ -32 & 128 \end{pmatrix}$

(d) $A^6 = \begin{pmatrix} 1 & 2 & 1 \\ 0 & -1 & -3 \\ 0 & 0 & 3 \end{pmatrix} \begin{pmatrix} 2 & 0 & 0 \\ 0 & 1 & 0 \\ 0 & 0 & -1 \end{pmatrix}^6 \begin{pmatrix} 1 & 2 & 5/3 \\ 0 & -1 & -1 \\ 0 & 0 & 1/3 \end{pmatrix}$

$$= \begin{pmatrix} 64 & 126 & 105 \\ 0 & 1 & 0 \\ 0 & 0 & 1 \end{pmatrix}$$

(e) $A^6 = \begin{pmatrix} 3 & 0 & 0 \\ 1 & 3 & -1 \\ 2 & 1 & 1 \end{pmatrix} \begin{pmatrix} 1 & 0 & 0 \\ 0 & 2 & 0 \\ 0 & 0 & -2 \end{pmatrix}^6 \begin{pmatrix} \frac{1}{3} & 0 & 0 \\ -\frac{1}{4} & \frac{1}{4} & \frac{1}{4} \\ -\frac{5}{12} & -\frac{1}{4} & \frac{3}{4} \end{pmatrix}$

$$= \begin{pmatrix} 1 & 0 & 0 \\ -21 & 64 & 0 \\ -42 & 0 & 64 \end{pmatrix}$$

(f) $$A^6 = \begin{pmatrix} 1 & 1 & -2 \\ 2 & 0 & 1 \\ 3 & 1 & 0 \end{pmatrix} \begin{pmatrix} 2 & 0 & 0 \\ 0 & 0 & 0 \\ 0 & 0 & 0 \end{pmatrix}^6 \begin{pmatrix} \frac{1}{2} & 1 & -\frac{1}{2} \\ -\frac{3}{2} & -3 & \frac{5}{2} \\ -1 & -1 & 1 \end{pmatrix}$$

$$= \begin{pmatrix} 32 & 64 & -32 \\ 64 & 128 & -64 \\ 96 & 192 & -96 \end{pmatrix}$$

3. If $A = XDX^{-1}$ is nonsingular, then $A^{-1} = XD^{-1}X^{-1}$

(a) $A^{-1} = XD^{-1}X^{-1} = XDX^{-1} = A$

(b) $$A^{-1} = \begin{pmatrix} -2 & -3 \\ 1 & 2 \end{pmatrix} \begin{pmatrix} \frac{1}{2} & 0 \\ 0 & 1 \end{pmatrix} \begin{pmatrix} -2 & -3 \\ 1 & 2 \end{pmatrix} = \begin{pmatrix} -1 & -3 \\ 1 & \frac{5}{2} \end{pmatrix}$$

(d) $$A^{-1} = \begin{pmatrix} 1 & 2 & 1 \\ 0 & -1 & -3 \\ 0 & 0 & 3 \end{pmatrix} \begin{pmatrix} 2 & 0 & 0 \\ 0 & 1 & 0 \\ 0 & 0 & -1 \end{pmatrix}^{-1} \begin{pmatrix} 1 & 2 & \frac{5}{3} \\ 0 & -1 & -1 \\ 0 & 0 & \frac{1}{3} \end{pmatrix}$$

$$= \begin{pmatrix} \frac{1}{2} & -1 & -\frac{3}{2} \\ 0 & 1 & 2 \\ 0 & 0 & -1 \end{pmatrix}$$

(e) $$A^{-1} = \begin{pmatrix} 3 & 0 & 0 \\ 1 & 3 & -1 \\ 2 & 1 & 1 \end{pmatrix} \begin{pmatrix} 1 & 0 & 0 \\ 0 & 2 & 0 \\ 0 & 0 & -2 \end{pmatrix}^{-1} \begin{pmatrix} \frac{1}{3} & 0 & 0 \\ -\frac{1}{4} & \frac{1}{4} & \frac{1}{4} \\ -\frac{5}{12} & -\frac{1}{4} & \frac{3}{4} \end{pmatrix}$$

$$= \begin{pmatrix} 1 & 0 & 0 \\ -\frac{1}{4} & \frac{1}{4} & \frac{3}{4} \\ \frac{3}{4} & \frac{1}{4} & -\frac{3}{4} \end{pmatrix}$$

4. (a) The eigenvalues of A are $\lambda_1 = 1$ and $\lambda_2 = 0$

$$A = XDX^{-1}$$

Since $D^2 = D$ it follows that

$$A^2 = XD^2X^{-1} = XDX^{-1} = A$$

(b) $$A = \begin{pmatrix} 1 & 1 & -1 \\ 0 & 1 & -1 \\ 0 & 0 & 1 \end{pmatrix} \begin{pmatrix} 9 & 0 & 0 \\ 0 & 4 & 0 \\ 0 & 0 & 1 \end{pmatrix} \begin{pmatrix} 1 & -1 & 0 \\ 0 & 1 & 1 \\ 0 & 0 & 1 \end{pmatrix}$$

$$B = XD^{1/2}X^{-1} = \begin{pmatrix} 1 & 1 & -1 \\ 0 & 1 & -1 \\ 0 & 0 & 1 \end{pmatrix} \begin{pmatrix} 3 & 0 & 0 \\ 0 & 2 & 0 \\ 0 & 0 & 1 \end{pmatrix} \begin{pmatrix} 1 & -1 & 0 \\ 0 & 1 & 1 \\ 0 & 0 & 1 \end{pmatrix}$$

$$= \begin{pmatrix} 3 & -1 & 1 \\ 0 & 2 & 1 \\ 0 & 0 & 1 \end{pmatrix}$$

5. If X diagonalizes A, then

$$X^{-1}AX = D$$

where D is a diagonal matrix. It follows that

$$D = D^T = X^T A^T (X^{-1})^T = Y^{-1} A^T Y$$

Therefore Y diagonalizes A^T.

6. If $A = XDX^{-1}$ where D is a diagonal matrix whose diagonal elements are all either 1 or -1, then $D^{-1} = D$ and

$$A^{-1} = XD^{-1}X^{-1} = XDX^{-1} = A$$

7. If $\mathbf{x}$ is an eigenvector belonging to the eigenvalue a, then

$$\begin{pmatrix} 0 & 1 & 0 \\ 0 & 0 & 1 \\ 0 & 0 & b-a \end{pmatrix} \begin{pmatrix} x_1 \\ x_2 \\ x_3 \end{pmatrix} = \begin{pmatrix} 0 \\ 0 \\ 0 \end{pmatrix}$$

and it follows that

$$x_2 = x_3 = 0$$

Thus the eigenspace corresponding to $\lambda_1 = \lambda_2 = a$ has dimension 1 and is spanned by $(1,\ 0,\ 0)^T$. The matrix is defective since a is a double eigenvalue and its eigenspace only has dimension 1.

8. (a) The characteristic polynomial of the matrix factors as follows.

$$p(\lambda) = \lambda(2 - \lambda)(\alpha - \lambda)$$

Thus the only way that the matrix can have a multiple eigenvalue is if $\alpha = 0$ or $\alpha = 2$. In the case $\alpha = 0$, we have that $\lambda = 0$ is an eigenvalue of multiplicity 2 and the corresponding eigenspace is spanned by the vectors $\mathbf{x}_1 = (-1, 1, 0)^T$ and $\mathbf{x}_2 = \mathbf{e}_3$. Since $\lambda = 0$ has two linearly independent eigenvectors, the matrix is not defective. Similarly in the case $\alpha = 2$ the matrix will not be defective since the eigenvalue $\lambda = 2$ possesses two linearly independent eigenvectors $\mathbf{x}_1 = (1, 1, 0)^T$ and $\mathbf{x}_2 = \mathbf{e}_3$.

9. If $A - \lambda I$ has rank 1, then

$$\dim N(A - \lambda I) = 4 - 1 = 3$$

Since λ has multiplicity 3 the matrix is not defective.

10. (a) The proof is by induction. In the case $m = 1$,

$$A\mathbf{x} = \sum_{i=1}^{n} \alpha_i A\mathbf{x}_i = \sum_{i=1}^{n} \alpha_i \lambda_i \mathbf{x}_i$$

If

$$A^k\mathbf{x} = \sum_{i=1}^{n} \alpha_i \lambda_i^k \mathbf{x}_i$$

then

$$A^{k+1}\mathbf{x} = A(A^k\mathbf{x}) = A\Big(\sum_{i=1}^{n} \alpha_i \lambda_i^k \mathbf{x}_i\Big) = \sum_{i=1}^{n} \alpha_i \lambda_i^k A\mathbf{x}_i = \sum_{i=1}^{n} \alpha_i \lambda_i^{k+1} \mathbf{x}_i$$

(b) If $\lambda_1 = 1$, then

$$A^m\mathbf{x} = \alpha_1\mathbf{x}_1 + \sum_{i=2}^{n} \alpha_i \lambda_i^m \mathbf{x}_i$$

Since $0 < \lambda_i < 1$ for $i = 2, \ldots, n$, it follows that $\lambda_i^m \to 0$ as $m \to \infty$. Hence

$$\lim_{m\to\infty} A^m\mathbf{x} = \alpha_1\mathbf{x}_1$$

11. If A is an $n \times n$ matrix and λ is an eigenvalue of multiplicity n then A is diagonalizable if and only if

$$\dim N(A - \lambda I) = n$$

or equivalently

$$\operatorname{rank}(A - \lambda I) = 0$$

The only way the rank can be 0 is if

$$\begin{aligned} A - \lambda I &= O \\ A &= \lambda I \end{aligned}$$

12. If A is nilpotent, then 0 is an eigenvalue of multiplicity n. It follows from Exercise 11 that A is diagonalizable if and only if $A = O$.

13. Let A be a diagonalizable $n \times n$ matrix. Let $\lambda_1, \lambda_2, \ldots, \lambda_k$ be the nonzero eigenvalues of A. The remaining eigenvalues are all 0.

$$\lambda_{k+1} = \lambda_{k+2} = \cdots = \lambda_n = 0$$

If $\mathbf{x}_i$ is an eigenvector belonging to λ_i, then

$$\begin{aligned} A\mathbf{x}_i &= \lambda_i\mathbf{x}_i \qquad && i = 1, \ldots, k \\ A\mathbf{x}_i &= \mathbf{0} && i = k+1, \ldots, n \end{aligned}$$

Since A is diagonalizable we can choose eigenvectors $\mathbf{x}_1, \ldots, \mathbf{x}_n$ which form a basis for R^n. Given any vector $\mathbf{x} \in R^n$ we can write

$$\mathbf{x} = c_1\mathbf{x}_1 + c_2\mathbf{x}_2 + \cdots + c_n\mathbf{x}_n$$

It follows that

$$A\mathbf{x} = c_1\lambda_1\mathbf{x}_1 + c_2\lambda_2\mathbf{x}_2 + \cdots + c_k\lambda_k\mathbf{x}_k$$

Thus $\mathbf{x}_1, \ldots, \mathbf{x}_k$ span the column space of A and since they are linearly independent they form a basis for the column space.

14. The matrix $\begin{pmatrix} 0 & 1 \\ 0 & 0 \end{pmatrix}$ has rank 1 even though all of its eigenvalues are 0.

15. (a) For $i = 1, \ldots, k$

$$\mathbf{b}_i = B\mathbf{e}_i = X^{-1}AX\mathbf{e}_i = X^{-1}A\mathbf{x}_i = \lambda X^{-1}\mathbf{x}_i = \lambda\mathbf{e}_i$$

Thus the first k columns of B will have λ's on the diagonal and 0's in the off diagonal positions.

(b) Clearly λ is an eigenvalue of B whose multiplicity is at least k. Since A and B are similar they have the same characteristic polynomial. Thus λ is an eigenvalue of A with multiplicity at least k.

16. (a) If $\mathbf{x}$ and $\mathbf{y}$ are nonzero vectors in R^n and $A = \mathbf{x}\mathbf{y}^T$, then A has rank 1. Thus

$$\dim N(A) = n - 1$$

It follows from Exercise 15 that $\lambda = 0$ is an eigenvalue with multiplicity greater than or equal to $n - 1$.

(b) By part (a)

$$\lambda_1 = \lambda_2 = \cdots = \lambda_{n-1} = 0$$

The sum of the eigenvalues is the trace of A which equals $\mathbf{x}^T\mathbf{y}$. Thus

$$\lambda_n = \sum_{i=1}^{n} \lambda_i = \operatorname{tr} A = \mathbf{x}^T\mathbf{y} = \mathbf{y}^T\mathbf{x}$$

Furthermore

$$A\mathbf{x} = \mathbf{x}\mathbf{y}^T\mathbf{x} = \lambda_n\mathbf{x}$$

so $\mathbf{x}$ is an eigenvector belonging to λ_n.

(c) Since $\dim N(A) = n-1$, it follows that $\lambda = 0$ has $n-1$ linearly independent eigenvectors $\mathbf{x}_1, \mathbf{x}_2, \ldots, \mathbf{x}_{n-1}$. If $\lambda_n \neq 0$ and $\mathbf{x}_n$ is an eigenvector belonging to λ_n, then $\mathbf{x}_n$ will be independent of $\mathbf{x}_1, \ldots, \mathbf{x}_{n-1}$ and hence A will have n linearly independent eigenvectors.

17. If A is diagonalizable, then

$$A = XDX^{-1}$$

where D is a diagonal matrix. If B is similar to A, then there exists a nonsingular matrix S such that $B = S^{-1}AS$. It follows that

$$\begin{aligned} B &= S^{-1}(XDX^{-1})S \\ &= (S^{-1}X)D(S^{-1}X)^{-1} \end{aligned}$$

Therefore B is diagonalizable with diagonalizing matrix $S^{-1}X$.

18. If $A = XD_1X^{-1}$ and $B = XD_2X^{-1}$, where D_1 and D_2 are diagonal matrices, then

$$\begin{aligned} AB &= (XD_1X^{-1})(XD_2X^{-1}) \\ &= XD_1D_2X^{-1} \end{aligned}$$

$$
\begin{aligned}
&= XD_2D_1X^{-1}\\
&= (XD_2X^{-1})(XD_1X^{-1})\\
&= BA
\end{aligned}
$$

19. If $\mathbf{r}_j$ is an eigenvector belonging $\lambda_j = t_{jj}$ then we claim that

$$r_{j+1,j} = r_{j+2,j} = \cdots + = r_{nj} = 0$$

The eigenvector $\mathbf{r}_j$ is a nontrivial solution to $(T - t_{jj}I)\mathbf{x} = \mathbf{0}$. The augmented matrix for this system is $(T - t_{jj}I \mid \mathbf{0})$. The equations corresponding to the last $n - j$ rows of the augmented matrix do not involve the variables $x_1, x_2, \ldots, x_j$. These last $n - j$ rows form a homogeneous system that is in strict triangular form with respect to the unknowns $x_{j+1}, x_{j+2}, \ldots, x_n$. The solution to this strictly triangular system is

$$x_{j+1} = x_{j+2} = \cdots = x_n = 0$$

Thus the last $n - j$ entries of the eigenvector $\mathbf{r}_j$ are all equal to 0. If we set $R = (\mathbf{r}_1, \mathbf{r}_2, \ldots, \mathbf{r}_n)$ then R is upper triangular and R diagonalizes T.

23. If A is stochastic then the entries of each of its column vectors will all add up to 1, so the entries of each of the row vectors of A^T will all add up to 1 and consequently $A^T\mathbf{e} = \mathbf{e}$. Therefore $\lambda = 1$ is an eigenvalue of A^T. Since A and A^T have the same eigenvalues, it follows that $\lambda = 1$ is an eigenvalue of A.

24. Since the rows of a doubly stochastic matrix A all add up to 1 it follows that $\mathbf{e}$ is an eigenvector of A belonging to the eigenvalue $\lambda = 1$. If $\lambda = 1$ is the dominant eigenvalue then for any starting probability vector $\mathbf{x}_0$, the Markov chain will converge to a steady-state vector $\mathbf{x} = c\mathbf{e}$. Since the steady-state vector must be a probability vector we have

$$1 = x_1 + x_2 + \cdots + x_n = c + c + \cdots + c = nc$$

and hence $c = \frac{1}{n}$.

25. Let

$$\mathbf{w}_k = M\mathbf{x}_k \qquad \text{and} \qquad \alpha_k = \frac{\mathbf{e}^T\mathbf{x}_k}{n}$$

It follows from equation (5) in the textbook that

$$\mathbf{x}_{k+1} = A\mathbf{x}_k = pM\mathbf{x}_k + \frac{1-p}{n}\mathbf{e}\mathbf{e}^T\mathbf{x}_k = p\mathbf{w}_k + (1-p)\alpha_k\mathbf{e}$$

26. (a) Since $A^2 = O$, it follows that

$$e^A = I + A = \begin{pmatrix} 2 & 1 \\ -1 & 0 \end{pmatrix}$$

(c) Since

$$A^k = \begin{pmatrix} 1 & 0 & -k \\ 0 & 1 & 0 \\ 0 & 0 & 1 \end{pmatrix} \qquad k = 1, 2, \ldots$$

it follows that

$$e^A = \begin{pmatrix} e & 0 & 1-e \\ 0 & e & 0 \\ 0 & 0 & e \end{pmatrix}$$

27. (b) $\begin{pmatrix} 2e - \frac{1}{e} & 2e - \frac{2}{e} \\ -e + \frac{1}{e} & -e + \frac{2}{e} \end{pmatrix}$

28. (d) The matrix A is defective, so e^{At} must be computed using the definition of the matrix exponential. Since

$$A^2 = \begin{pmatrix} 1 & 0 & 1 \\ 0 & 0 & 0 \\ -1 & 0 & -1 \end{pmatrix} \quad \text{and} \quad A^3 = O$$

it follows that

$$\begin{aligned} e^{At} &= I + tA + \frac{t^2}{2}A^2 \\ &= \begin{pmatrix} 1+t+\frac{1}{2}t^2 & t & t+\frac{1}{2}t^2 \\ t & 1 & t \\ -t-\frac{1}{2}t^2 & -t & 1-t-\frac{1}{2}t^2 \end{pmatrix} \end{aligned}$$

The solution to the initial value problem is

$$\mathbf{Y} = e^{At}\mathbf{Y}_0 = \begin{pmatrix} 1+t \\ 1 \\ -1-t \end{pmatrix}$$

29. If λ is an eigenvalue of A and $\mathbf{x}$ is an eigenvector belonging to λ then

$$\begin{aligned} e^A\mathbf{x} &= \left(I + A + \frac{1}{2!}A^2 + \frac{1}{3!}A^3 + \cdots\right)\mathbf{x} \\ &= \mathbf{x} + A\mathbf{x} + \frac{1}{2!}A^2\mathbf{x} + \frac{1}{3!}A^3\mathbf{x} + \cdots \\ &= \mathbf{x} + \lambda\mathbf{x} + \frac{1}{2!}\lambda^2\mathbf{x} + \frac{1}{3!}\lambda^3\mathbf{x} + \cdots \\ &= \left(1 + \lambda + \frac{1}{2!}\lambda^2 + \frac{1}{3!}\lambda^3 + \cdots\right)\mathbf{x} \\ &= e^\lambda\mathbf{x} \end{aligned}$$

30. If A is diagonalizable with linearly independent eigenvectors $\mathbf{x}_1, \ldots, \mathbf{x}_n$ then, by Exercise 29, $\mathbf{x}_1, \ldots, \mathbf{x}_n$ are eigenvectors of e^A. Furthermore, if $\mathbf{x}_1, \ldots, \mathbf{x}_k$ are eigenvectors corresponding to the eigenvalue λ of A and the eigenvalue e^λ of e^A, then these eigenvalues must have multiplicity at least k (see Exercise 15). Thus if $\lambda_1, \ldots, \lambda_n$ are the eigenvalues of A, then $e^{\lambda_1}, \ldots, e^{\lambda_n}$ are the eigenvalues of e^A. Since the eigenvalues of e^A are all nonzero, e^A is nonsingular.

31. (a) Let A be a diagonalizable matrix with characteristic polynomial

$$p(\lambda) = a_1\lambda^n + a_2\lambda^{n-1} + \cdots + a_n\lambda + a_{n+1}$$

and let D be a diagonal matrix whose diagonal entries $\lambda_1, \ldots, \lambda_n$ are the eigenvalues of A. The matrix

$$p(D) = a_1 D^n + a_2 D^{n-1} + \cdots + a_n D + a_{n+1} I$$

is diagonal since it is a sum of diagonal matrices. Furthermore the jth diagonal entry of $p(D)$ is

$$a_1 \lambda_j^n + a_2 \lambda_j^{n-1} + \cdots + a_n \lambda_j + a_{n+1} = p(\lambda_j) = 0$$

Therefore $p(D) = O$.

(b) If $A = XDX^{-1}$, then

$$\begin{aligned} p(A) &= a_1 A^n + a_2 A^{n-1} + \cdots + a_n A + a_{n+1} I \\ &= a_1 X D^n X^{-1} + a_2 X D^{n-1} X^{-1} + \cdots + a_n X D X^{-1} + a_{n+1} X I X^{-1} \\ &= X(a_1 D^n + a_2 D^{n-1} + \cdots + a_n D + a_{n+1}) X^{-1} \\ &= X p(D) X^{-1} \\ &= O \end{aligned}$$

(c) In part (b) we showed that

$$p(A) = a_1 A^n + a_2 A^{n-1} + \cdots + a_n A + a_{n+1} I = O$$

If $a_{n+1} \neq 0$, then we can solve for I.

$$I = c_1 A^n + c_2 A^{n-1} + \cdots + c_n A$$

where $c_j = -\dfrac{a_j}{a_{n+1}}$ for $j = 1, \ldots, n$. Thus if we set

$$q(A) = c_1 A^{n-1} + c_2 A^{n-2} + \cdots + c_{n-1} A + c_n I$$

then

$$I = Aq(A)$$

and it follows that A is nonsingular and

$$A^{-1} = q(A)$$

SECTION 4

2. (a) $\mathbf{z}_2^H \mathbf{z}_1 = \begin{pmatrix} \dfrac{-i}{\sqrt{2}} & -\dfrac{1}{\sqrt{2}} \end{pmatrix} \begin{pmatrix} \dfrac{1+i}{2} \\ \dfrac{1-i}{2} \end{pmatrix} = 0$

$$\mathbf{z}_1^H \mathbf{z}_1 = \begin{pmatrix} \dfrac{1-i}{2} & \dfrac{1+i}{2} \end{pmatrix} \begin{pmatrix} \dfrac{1+i}{2} \\ \dfrac{1-i}{2} \end{pmatrix} = 1$$

$$\mathbf{z}_2^H \mathbf{z}_2 = \begin{pmatrix} \frac{-i}{\sqrt{2}} & -\frac{1}{\sqrt{2}} \end{pmatrix} \begin{pmatrix} \frac{i}{\sqrt{2}} \\ -\frac{1}{\sqrt{2}} \end{pmatrix} = 1$$

5. There will not be a unique unitary diagonalizing matrix for a given Hermitian matrix A, however, the column vectors of any unitary diagonalizing matrix must be unit eigenvectors of A.

(a) $\lambda_1 = 3$ has a unit eigenvector $\left(\frac{1}{\sqrt{2}}, \frac{1}{\sqrt{2}}\right)^T$

$\lambda_2 = 1$ has a unit eigenvector $\left(\frac{1}{\sqrt{2}}, -\frac{1}{\sqrt{2}}\right)^T$.

$$Q = \frac{1}{\sqrt{2}} \begin{pmatrix} 1 & 1 \\ 1 & -1 \end{pmatrix}$$

(b) $\lambda_1 = 6$ has a unit eigenvector $\left(\frac{2}{\sqrt{14}}, \frac{3-i}{\sqrt{14}}\right)^T$

$\lambda_2 = 1$ has a unit eigenvector $\left(\frac{-5}{\sqrt{35}}, \frac{3-i}{\sqrt{25}}\right)^T$

$$Q = \begin{pmatrix} \frac{2}{\sqrt{14}} & -\frac{5}{\sqrt{35}} \\ \frac{3-i}{\sqrt{14}} & \frac{3-i}{\sqrt{35}} \end{pmatrix}$$

(c) $\lambda_1 = 3$ has a unit eigenvector $\left(-\frac{1}{\sqrt{2}}, \frac{i}{\sqrt{2}}, 0\right)^T$

$\lambda_2 = 2$ has a unit eigenvector $(0,\ 0,\ 1)^T$

$\lambda_3 = 1$ has a unit eigenvector $\left(\frac{1}{\sqrt{2}}, \frac{i}{\sqrt{2}}, 0\right)^T$

$$Q = \frac{1}{\sqrt{2}} \begin{pmatrix} -1 & 0 & 1 \\ i & 0 & i \\ 0 & \sqrt{2} & 0 \end{pmatrix}$$

(d) $\lambda_1 = 5$ has a unit eigenvector $\left(0, \frac{1}{\sqrt{2}}, -\frac{1}{\sqrt{2}}\right)^T$

$\lambda_2 = 3$ has a unit eigenvector $\left(\frac{2}{\sqrt{6}}, \frac{1}{\sqrt{6}}, \frac{1}{\sqrt{6}}\right)^T$

$\lambda_3 = 0$ has a unit eigenvector $\left(-\frac{1}{\sqrt{3}}, \frac{1}{\sqrt{3}}, \frac{1}{\sqrt{3}}\right)^T$

$$Q = \frac{1}{\sqrt{6}} \begin{pmatrix} 0 & 2 & -\sqrt{2} \\ \sqrt{3} & 1 & \sqrt{2} \\ -\sqrt{3} & 1 & \sqrt{2} \end{pmatrix}$$

(e) The eigenvalue $\lambda_1 = -1$ has unit eigenvector $\frac{1}{\sqrt{2}}(-1,\ 0,\ 1)^T$.
The eigenvalues $\lambda_2 = \lambda_3 = 1$ have unit eigenvectors $\frac{1}{\sqrt{2}}(1,\ 0,\ 1)^T$ and $(0,\ 1,\ 0)^T$. The three vectors form an orthonormal set. Thus

$$Q = \begin{pmatrix} -\frac{1}{\sqrt{2}} & \frac{1}{\sqrt{2}} & 0 \\ 0 & 0 & 1 \\ \frac{1}{\sqrt{2}} & \frac{1}{\sqrt{2}} & 0 \end{pmatrix}$$

is an orthogonal diagonalizing matrix.

(f) $\lambda_1 = 3$ has a unit eigenvector $\mathbf{q}_1 = \left(\frac{1}{\sqrt{3}}, \frac{1}{\sqrt{3}}, \frac{1}{\sqrt{3}}\right)^T$.
$\lambda_2 = \lambda_3 = 0$. The eigenspace corresponding to $\lambda = 0$ has dimension 2. It consists of all vectors $\mathbf{x}$ such that

$$x_1 + x_2 + x_3 = 0$$

In this case we must choose a basis for the eigenspace consisting of orthogonal unit vectors. If we take $\mathbf{q}_2 = \frac{1}{\sqrt{2}}(-1,\ 0,\ 1)^T$ and $\mathbf{q}_3 = \frac{1}{\sqrt{6}}(-1,\ 2,\ -1)^T$ then

$$Q = \frac{1}{\sqrt{6}} \begin{pmatrix} \sqrt{2} & -\sqrt{3} & -1 \\ \sqrt{2} & 0 & 2 \\ \sqrt{2} & \sqrt{3} & -1 \end{pmatrix}$$

(g) $\lambda_1 = 6$ has unit eigenvector $\frac{1}{\sqrt{6}}(-2,\ -1,\ 1)^T$, $\lambda_2 = \lambda_3 = 0$. The vectors $\mathbf{x}_1 = (1,\ 0,\ 2)^T$ and $\mathbf{x}_2 = (-1,\ 2,\ 0)^T$ form a basis for the eigenspace corresponding to $\lambda = 0$. The Gram–Schmidt process can be used to construct an orthonormal basis.

$$\begin{aligned}
r_{11} &= \|\mathbf{x}_1\| = \sqrt{5} \\
\mathbf{q}_1 &= \frac{1}{\sqrt{5}}\mathbf{x}_1 = \frac{1}{\sqrt{5}}(1,\ 0,\ 2)^T \\
\mathbf{p}_1 &= (\mathbf{x}_2^T\mathbf{q}_1)\mathbf{q}_1 = -\frac{1}{\sqrt{5}}\mathbf{q}_1 = -\frac{1}{5}(1,\ 0,\ 2)^T \\
\mathbf{x}_2 - \mathbf{p}_1 &= \left(-\frac{4}{5},\ 2,\ \frac{2}{5}\right)^T \\
r_{22} &= \|\mathbf{x}_2 - \mathbf{p}_1\| = \frac{2\sqrt{30}}{5} \\
\mathbf{q}_2 &= \frac{1}{\sqrt{30}}(-2,\ 5,\ 1)^T
\end{aligned}$$

Thus

$$Q = \begin{pmatrix} \frac{1}{\sqrt{5}} & -\frac{2}{\sqrt{30}} & -\frac{2}{\sqrt{6}} \\ 0 & \frac{5}{\sqrt{30}} & -\frac{1}{\sqrt{6}} \\ \frac{2}{\sqrt{5}} & \frac{1}{\sqrt{30}} & \frac{1}{\sqrt{6}} \end{pmatrix}$$

6. If A is Hermitian, then $A^H = A$. Comparing the diagonal entries of A^H and A we see that

$$\overline{a}_{ii} = a_{ii} \quad \text{for} \quad i = 1, \ldots, n$$

Thus if A is Hermitian, then its diagonal entries must be real.

7. (a)

$$(A^H)^H = \overline{\left(\overline{A}^T\right)}^T = \left(\overline{\overline{A}}^T\right)^T = A$$

(b)

$$(\alpha A + \beta C)^H = \overline{\alpha A + \beta C}^T = (\overline{\alpha}\,\overline{A} + \overline{\beta}\,\overline{C})^T = \overline{\alpha}\,\overline{A}^T + \overline{\beta}\,\overline{C}^T = \overline{\alpha}A^H + \overline{\beta}C^H$$

(c) In general

$$\overline{AB} = \overline{A}\,\overline{B}$$

(See Exercise 17 of Section 1.) Using this we have

$$(AB)^H = (\overline{AB})^T = (\overline{A}\,\overline{B})^T = \overline{B}^T\overline{A}^T = B^H A^H$$

8. (i) $\langle \mathbf{z}, \mathbf{z} \rangle = \mathbf{z}^H\mathbf{z} = \Sigma |z_i|^2 \ge 0$ with equality if and only if $\mathbf{z} = \mathbf{0}$

(ii) $\overline{\langle \mathbf{w}, \mathbf{z} \rangle} = \overline{\mathbf{z}^H\mathbf{w}} = \mathbf{z}^T\overline{\mathbf{w}} = \overline{\mathbf{w}}^T\mathbf{z} = \mathbf{w}^H\mathbf{z} = \langle \mathbf{z}, \mathbf{w} \rangle$

(iii)
$$\begin{aligned} \langle \alpha\mathbf{z} + \beta\mathbf{w}, \mathbf{u} \rangle &= \mathbf{u}^H(\alpha\mathbf{z} + \beta\mathbf{w}) \\ &= \alpha\mathbf{u}^H\mathbf{z} + \beta\mathbf{u}^H\mathbf{w} \\ &= \alpha\langle \mathbf{z}, \mathbf{u} \rangle + \beta\langle \mathbf{w}, \mathbf{u} \rangle \end{aligned}$$

9.

$$\begin{aligned} \langle \mathbf{z}, \alpha\mathbf{x} + \beta\mathbf{y} \rangle &= \overline{\langle \alpha\mathbf{x} + \beta\mathbf{y}, \mathbf{z} \rangle} \\ &= \overline{\alpha\,\langle \mathbf{x}, \mathbf{z} \rangle + \beta\,\langle \mathbf{y}, \mathbf{z} \rangle} \\ &= \overline{\alpha}\,\overline{\langle \mathbf{x}, \mathbf{z} \rangle} + \overline{\beta}\,\overline{\langle \mathbf{y}, \mathbf{z} \rangle} \\ &= \overline{\alpha}\,\langle \mathbf{z}, \mathbf{x} \rangle + \overline{\beta}\,\langle \mathbf{z}, \mathbf{y} \rangle \end{aligned}$$

10. For $j = 1, \ldots, n$

$$\langle \mathbf{z}, \mathbf{u}_j \rangle = \langle a_1\mathbf{u}_1 + \cdots + a_n\mathbf{u}_n, \mathbf{u}_j \rangle = a_1\,\langle \mathbf{u}_1, \mathbf{u}_j \rangle + \cdots + a_n\,\langle \mathbf{u}_n, \mathbf{u}_j \rangle = a_j$$

Using the result from Exercise 9 we have

$$\begin{aligned} \langle \mathbf{z}, \mathbf{w} \rangle &= \langle \mathbf{z}, b_1\mathbf{u}_1 + \cdots + b_n\mathbf{u}_n \rangle \\ &= \overline{b_1}\,\langle \mathbf{z}, \mathbf{u}_1 \rangle + \cdots + \overline{b_n}\,\langle \mathbf{z}, \mathbf{u}_n \rangle \\ &= \overline{b_1}\,a_1 + \cdots + \overline{b_n}\,a_n \end{aligned}$$

11. The matrix A can be factored into a product $A = QDQ^H$ where

$$Q = \frac{1}{\sqrt{2}}\begin{pmatrix} \sqrt{2} & 0 & 0 \\ 0 & i & -i \\ 0 & 1 & 1 \end{pmatrix} \quad \text{and} \quad D = \begin{pmatrix} 4 & 0 & 0 \\ 0 & 2 & 0 \\ 0 & 0 & 0 \end{pmatrix}$$

Let

$$E = \begin{pmatrix} 2 & 0 & 0 \\ 0 & \sqrt{2} & 0 \\ 0 & 0 & 0 \end{pmatrix}$$

Note that $E^H E = D$. If we set

$$B = EQ^H = \begin{pmatrix} 2 & 0 & 0 \\ 0 & -i & 1 \\ 0 & 0 & 0 \end{pmatrix}$$

then

$$B^H B = (EQ^H)^H (EQ^H) = QE^H EQ^H = QDQ^H = A$$

12. (a) $U^H U = I = UU^H$

(c) If $\mathbf{x}$ is an eigenvector belonging to λ then

$$\|\mathbf{x}\| = \|U\mathbf{x}\| = \|\lambda\mathbf{x}\| = |\lambda|\,\|\mathbf{x}\|$$

Therefore $|\lambda|$ must equal 1.

14. Let U be a matrix that is both unitary and Hermitian. If λ is an eigenvalue of U and $\mathbf{z}$ is an eigenvector belonging to λ, then

$$U^2\mathbf{z} = U^H U\mathbf{z} = I\mathbf{z} = \mathbf{z}$$

and

$$U^2\mathbf{z} = U(U\mathbf{z}) = U(\lambda\mathbf{z}) = \lambda(U\mathbf{z}) = \lambda^2\mathbf{z}$$

Therefore

$$\begin{aligned} \mathbf{z} &= \lambda^2\mathbf{z} \\ (1-\lambda^2)\mathbf{z} &= \mathbf{0} \end{aligned}$$

Since $\mathbf{z} \neq \mathbf{0}$ it follows that $\lambda^2 = 1$.

15. (a) A and T are similar and hence have the same eigenvalues. Since T is triangular, its eigenvalues are t_{11} and t_{22}.

(b) It follows from the Schur decomposition of A that

$$AU = UT$$

where U is unitary. Comparing the first columns of each side of this equation we see that

$$A\mathbf{u}_1 = U\mathbf{t}_1 = t_{11}\mathbf{u}_1$$

Hence $\mathbf{u}_1$ is an eigenvector belonging to t_{11}.

(c) Comparing the second column of $AU = UT$, we see that

$$\begin{aligned} A\mathbf{u}_2 &= U\mathbf{t}_2 \\ &= t_{12}\mathbf{u}_1 + t_{22}\mathbf{u}_2 \\ &\neq t_{22}\mathbf{u}_2 \end{aligned}$$

since $t_{12}\mathbf{u}_1 \neq \mathbf{0}$.

16. If A has Schur decomposition UTU^H and the diagonal entries of T are all distinct then by Exercise 19 in Section 3 there is an upper triangular matrix R that diagonalizes T. Thus we can factor T into a product RDR^{-1} where D is a diagonal matrix. It follows that

$$A = UTU^H = U(RDR^{-1})U^H = (UR)D(R^{-1}U^H)$$

and hence the matrix $X = UR$ diagonalizes A.

17. $M^H = (A - iB)^T = A^T - iB^T$
$-M = -A - iB$
Therefore $M^H = -M$ if and only if $A^T = -A$ and $B^T = B$.

18. If A is skew Hermitian, then $A^H = -A$. Let λ be any eigenvalue of A and let $\mathbf{z}$ be a unit eigenvector belonging to λ. It follows that

$$\mathbf{z}^H A\mathbf{z} = \lambda\mathbf{z}^H\mathbf{z} = \lambda\|\mathbf{z}\|^2 = \lambda$$

and hence

$$\overline{\lambda} = \lambda^H = (\mathbf{z}^H A\mathbf{z})^H = \mathbf{z}^H A^H\mathbf{z} = -\mathbf{z}^H A\mathbf{z} = -\lambda$$

This implies that λ is purely imaginary.

19. If A is normal then there exists a unitary matrix U that diagonalizes A. If D is the diagonal matrix whose diagonal entries are the eigenvalues of A then $A = UDU^H$. The column vectors of U are orthonormal eigenvectors of A.

(a) Since $A^H = (UDU^H)^H = UD^HU^H$ and the matrix D^H is diagonal, we have that U diagonalizes A^H. Therefore A^H has a complete orthonormal set of eigenvectors and hence it is a normal matrix.

(b) $I + A = I + UDU^H = UIU^H + +UDU^H = +U(I + D)U^H$.
The matrix $I + D$ is diagonal, so U diagonalizes $I + A$. Therefore $I + A$ has a complete orthonormal set of eigenvectors and hence it is a normal matrix.

(c) $A^2 = UD^2U^H$.
The matrix D^2 is diagonal, so U diagonalizes A^2. Therefore A^2 has a complete orthonormal set of eigenvectors and hence it is a normal matrix.

20. $B = SAS^{-1} = \begin{pmatrix} a_{11} & \sqrt{a_{12}a_{21}} \\ \sqrt{a_{12}a_{21}} & a_{22} \end{pmatrix}$

Since B is symmetric it has real eigenvalues and an orthonormal set of eigenvectors. The matrix A is similar to B, so it has the same eigenvalues. Indeed, A is similar to the diagonal matrix D whose diagonal entries are the eigenvalues of B. Therefore A is diagonalizable and hence it has two linearly independent eigenvectors.

21. (a) $A^{-1} = \begin{pmatrix} 1 & 1-c & -1-c \\ 1 & 2 & 1 \\ 0 & 1 & 1 \end{pmatrix}$

$$A^{-1}CA = \begin{pmatrix} 0 & 1 & 0 \\ 1 & c+1 & 1 \\ 0 & 1 & -1 \end{pmatrix}$$

(b) Let $B = A^{-1}CA$. Since B and C are similar they have the same eigenvalues. The eigenvalues of C are the roots of $p(x)$. Thus the roots of $p(x)$ are the eigenvalues of B. We saw in part (a) that B is symmetric. Thus all of the eigenvalues of B are real.

22. If A is Hermitian, then there is a unitary U that diagonalizes A. Thus

$$\begin{aligned} A &= UDU^H \\ &= (\mathbf{u}_1, \mathbf{u}_2, \ldots, \mathbf{u}_n) \begin{pmatrix} \lambda_1 & & & \\ & \lambda_2 & & \\ & & \ddots & \\ & & & \lambda_n \end{pmatrix} \begin{pmatrix} \mathbf{u}_1^H \\ \mathbf{u}_2^H \\ \vdots \\ \mathbf{u}_n^H \end{pmatrix} \\ &= (\lambda_1\mathbf{u}_1, \lambda_2\mathbf{u}_2, \ldots, \lambda_n\mathbf{u}_n) \begin{pmatrix} \mathbf{u}_1^H \\ \mathbf{u}_2^H \\ \vdots \\ \mathbf{u}_n^H \end{pmatrix} \\ &= \lambda_1\mathbf{u}_1\mathbf{u}_1^H + \lambda_2\mathbf{u}_2\mathbf{u}_2^H + \cdots + \lambda_n\mathbf{u}_n\mathbf{u}_n^H \end{aligned}$$

24. (a) Since the eigenvectors $\mathbf{u}_1, \ldots, \mathbf{u}_n$ form an orthonormal basis for C^n, the coordinates of $\mathbf{x}$ with respect to this basis are $c_i = \mathbf{u}_i^H\mathbf{x}_i$ for $i = 1, \ldots, n$. It follows then that

$$\begin{aligned} \mathbf{x} &= c_1\mathbf{u}_1 + c_2\mathbf{u}_2 + \cdots + c_n\mathbf{u}_n \\ A\mathbf{x} &= c_1A\mathbf{u}_1 + c_2A\mathbf{u}_2 + \cdots + c_nA\mathbf{u}_n \\ &= \lambda_1c_1\mathbf{u}_1 + \lambda_2c_2\mathbf{u}_2 + \cdots + \lambda_nc_n\mathbf{u}_n \\ \mathbf{x}^HA\mathbf{x} &= \lambda_1c_1\mathbf{x}^H\mathbf{u}_1 + \lambda_2c_2\mathbf{x}^H\mathbf{u}_2 + \cdots + \lambda_nc_n\mathbf{x}^H\mathbf{u}_n \\ &= \lambda_1c_1\bar{c}_1 + \lambda_2c_2\bar{c}_2 + \cdots + \lambda_nc_n\bar{c}_n \\ &= \lambda_1|c_1|^2 + \lambda_2|c_2|^2 + \cdots + \lambda_n|c_n|^2 \end{aligned}$$

By Parseval's formula

$$\mathbf{x}^H\mathbf{x} = \|\mathbf{x}\|^2 = \|\mathbf{c}\|^2$$

Thus

$$\begin{aligned}\rho(\mathbf{x}) &= \frac{\mathbf{x}^H A\mathbf{x}}{\mathbf{x}^H\mathbf{x}} \\ &= \frac{\lambda_1|c_1|^2 + \lambda_2|c_2|^2 + \cdots + \lambda_n|c_n|^2}{\|\mathbf{c}\|^2}\end{aligned}$$

(b) It follows from part (a) that

$$\frac{\lambda_{\min}\sum_{i=1}^{n}|c_i|^2}{\|\mathbf{c}\|^2} \le \rho(\mathbf{x}) \le \frac{\lambda_{\max}\sum_{i=1}^{n}|c_i|^2}{\|\mathbf{c}\|^2}$$

$$\lambda_{\min} \le \rho(\mathbf{x}) \le \lambda_{\max}$$

SECTION 5

1. If A has singular value decomposition $U\Sigma V^T$, then A^T has singular value decomposition $V\Sigma^T U^T$. The matrices Σ and Σ^T will have the same nonzero diagonal elements. Thus A and A^T have the same nonzero singular values.

3. If A is a matrix with singular value decomposition $U\Sigma V^T$, then the rank of A is the number of nonzero singular values it possesses, the 2-norm is equal to its largest singular value, and the closest matrix of rank 1 is $\sigma_1\mathbf{u}_1\mathbf{v}_1^T$.

(a) The rank of A is 1 and $\|A\|_2 = \sqrt{10}$. The closest matrix of rank 1 is A itself.

(c) The rank of A is 2 and $\|A\|_2 = 4$. The closest matrix of rank 1 is given by

$$4\mathbf{u}_1\mathbf{v}_1 = \begin{pmatrix} 2 & 2 \\ 2 & 2 \\ 0 & 0 \\ 0 & 0 \end{pmatrix}$$

(d) The rank of A is 3 and $\|A\|_2 = 3$. The closest matrix of rank 1 is given by

$$3\mathbf{u}_1\mathbf{v}_1 = \begin{pmatrix} 0 & 0 & 0 \\ 0 & \frac{3}{2} & \frac{3}{2} \\ 0 & \frac{3}{2} & \frac{3}{2} \\ 0 & 0 & 0 \end{pmatrix}$$

5. (b) Basis for $R(A)$: $\mathbf{u}_1 = \left(\frac{1}{2}, \frac{1}{2}, \frac{1}{2}, \frac{1}{2}\right)^T$, $\mathbf{u}_2 = \left(\frac{1}{2}, -\frac{1}{2}, -\frac{1}{2}, \frac{1}{2}\right)^T$

Basis for $N(A^T)$: $\mathbf{u}_3 = \left(\frac{1}{2}, -\frac{1}{2}, \frac{1}{2}, -\frac{1}{2}\right)$, $\mathbf{u}_4 = \left(\frac{1}{2}, \frac{1}{2}, -\frac{1}{2}, -\frac{1}{2}\right)^T$

6. If A is symmetric then $A^TA = A^2$. Thus the eigenvalues of A^TA are $\lambda_1^2, \lambda_2^2, \ldots, \lambda_n^2$. The singular values of A are the positive square roots of the eigenvalues of A^TA.

7. The vectors $\mathbf{v}_{r+1}, \ldots, \mathbf{v}_n$ are all eigenvectors belonging to $\lambda = 0$. Hence these vectors are all in $N(A)$ and since dim $N(A) = n - r$, they form a basis for $N(A)$. The vectors $\mathbf{v}_1, \ldots, \mathbf{v}_r$ are all vectors in $N(A)^\perp = R(A^T)$. Since dim $R(A^T) = r$, it follows that $\mathbf{v}_1, \ldots, \mathbf{v}_r$ form an orthonormal basis for $R(A^T)$.

8. If A is an $n \times n$ matrix with singular value decomposition $A = U\Sigma V^T$, then

$$A^TA = V\Sigma^2V^T \qquad \text{and} \qquad AA^T = U\Sigma^2U^T$$

If we set $X = VU^T$ then X is nonsingular and

$$X^{-1}(A^TA)X = UV^TV\Sigma^2V^TVU^T = U\Sigma^2U^T = AA^T$$

Therefore A^TA and AA^T are similar.

9. If σ is a singular value of A, then σ^2 is an eigenvalue of A^TA. Let $\mathbf{x}$ be an eigenvector of A^TA belonging to σ^2. It follows that

$$A^TA\mathbf{x} = \sigma^2\mathbf{x}$$

$$\mathbf{x}^TA^TA\mathbf{x} = \sigma^2\mathbf{x}^T\mathbf{x}$$

$$\|A\mathbf{x}\|_2^2 = \sigma^2\|\mathbf{x}\|_2^2$$

$$\sigma = \frac{\|A\mathbf{x}\|_2}{\|\mathbf{x}\|_2}$$

10. $$\begin{aligned} A^TA\hat{\mathbf{x}} &= A^TAA^+\mathbf{b} \\ &= V\Sigma^TU^TU\Sigma V^TV\Sigma^+U^T\mathbf{b} \\ &= V\Sigma^T\Sigma\Sigma^+U^T\mathbf{b} \end{aligned}$$

For any vector $\mathbf{y} \in R^m$

$$\Sigma^T\Sigma\Sigma^+\mathbf{y} = (\sigma_1y_1, \sigma_2y_2, \ldots, \sigma_ny_n)^T = \Sigma^T\mathbf{y}$$

Thus

$$A^TA\hat{\mathbf{x}} = V\Sigma^T\Sigma\Sigma^+(U^T\mathbf{b}) = V\Sigma^TU^T\mathbf{b} = A^T\mathbf{b}$$

11. $P = AA^+ = U\Sigma V^TV\Sigma^+U^T = U\Sigma\Sigma^+U^T$

The matrix $\Sigma\Sigma^+$ is an $m \times m$ diagonal matrix whose diagonal entries are all 0's and 1's. Thus we have

$$(\Sigma\Sigma^+)^T = \Sigma\Sigma^+ \quad \text{and} \quad (\Sigma\Sigma^+)^2 = \Sigma\Sigma^+$$

and it follows that

$$\begin{aligned} P^2 &= U(\Sigma\Sigma^+)^2U^T = U\Sigma^+\Sigma U^T = P \\ P^T &= U(\Sigma\Sigma^+)^TU^T = U\Sigma^+\Sigma U^T = P \end{aligned}$$

SECTION 6

1. (c) $\begin{pmatrix} 1 & 1/2 & -1 \\ 1/2 & 2 & 3/2 \\ -1 & 3/2 & 1 \end{pmatrix}$

2. $\lambda_1 = 4$, $\lambda_2 - 2$

$$Q = \begin{pmatrix} \frac{1}{\sqrt{2}} & -\frac{1}{\sqrt{2}} \\ \frac{1}{\sqrt{2}} & \frac{1}{\sqrt{2}} \end{pmatrix}$$

If we set

$$\begin{pmatrix} x \\ y \end{pmatrix} = Q \begin{pmatrix} x' \\ y' \end{pmatrix}$$

then

$$(x \ \ y)A \begin{pmatrix} x \\ y \end{pmatrix} = (x' \ \ y')Q^T AQ \begin{pmatrix} x' \\ y' \end{pmatrix}$$

It follows that

$$Q^T AQ = \begin{pmatrix} 4 & 0 \\ 0 & 2 \end{pmatrix}$$

and the equation of the conic can be written in the form

$$4(x')^2 + 2(y')^2 = 8$$
$$\frac{(x')^2}{2} + \frac{(y')^2}{4} = 1$$

The positive x' axis will be in the first quadrant in the direction of

$$\mathbf{q}_1 = \left(\frac{1}{\sqrt{2}}, \frac{1}{\sqrt{2}} \right)^T$$

The positive y' axis will be in the second quadrant in the direction of

$$\mathbf{q}_2 = \left(-\frac{1}{\sqrt{2}}, \frac{1}{\sqrt{2}} \right)^T$$

The graph will be exactly the same as Figure 6.6.3 except for the labeling of the axes.

3. (b) $A = \begin{pmatrix} 3 & 4 \\ 4 & 3 \end{pmatrix}$. The eigenvalues are $\lambda_1 = 7$, $\lambda_2 = -1$ with orthonormal eigenvectors

$$\left(\frac{1}{\sqrt{2}}, \frac{1}{\sqrt{2}} \right)^T \quad \text{and} \quad \left(-\frac{1}{\sqrt{2}}, \frac{1}{\sqrt{2}} \right)^T \quad \text{respectively.}$$

Let

$$Q = \frac{1}{\sqrt{2}} \begin{pmatrix} 1 & -1 \\ 1 & 1 \end{pmatrix} \quad \text{and} \quad \begin{pmatrix} x' \\ y' \end{pmatrix} = Q^T \begin{pmatrix} x \\ y \end{pmatrix}$$

The equation simplifies to

$$7(x')^2 - (y')^2 = -28$$

$$\frac{(y')^2}{28} - \frac{(x')^2}{4} = 1$$

which is in standard form with respect to the $x'y'$ axis system.

(c) $A = \begin{pmatrix} -3 & 3 \\ 3 & 5 \end{pmatrix}.$

The eigenvalues are $\lambda_1 = 6$, $\lambda_2 = -4$ with orthonormal eigenvectors

$$\left(\frac{1}{\sqrt{10}}, \frac{3}{\sqrt{10}}\right)^T \quad \text{and} \quad \left(-\frac{3}{\sqrt{10}}, \frac{1}{\sqrt{10}}\right)^T, \quad \text{respectively.}$$

Let

$$Q = \frac{1}{\sqrt{10}}\begin{pmatrix} 1 & -3 \\ 3 & 1 \end{pmatrix} \quad \text{and} \quad \begin{pmatrix} x' \\ y' \end{pmatrix} = Q^T \begin{pmatrix} x \\ y \end{pmatrix}$$

The equation simplifies to

$$6(x')^2 - 4(y')^2 = 24$$

$$\frac{(x')^2}{4} - \frac{(y')^2}{6} = 1$$

4. Using a suitable rotation of axes, the equation translates to

$$\lambda_1(x')^2 + \lambda_2(y')^2 = 1$$

Since λ_1 and λ_2 differ in sign, the graph will be an hyperbola.

5. The equation can be transformed into the form

$$\lambda_1(x')^2 + \lambda_2(y')^2 = \alpha$$

If either λ_1 and λ_2 is 0, then the graph is a pair of lines. Thus the conic section will be nondegenerate if and only if the eigenvalues of A are nonzero. The eigenvalues of A will be nonzero if and only if A is nonsingular.

6. (c) The eigenvalues are $\lambda_1 = 5$, $\lambda_2 = 2$. Therefore the matrix is positive definite.

(f) The eigenvalues are $\lambda_1 = 8$, $\lambda_2 = 2$, $\lambda_3 = 2$. Since all of the eigenvalues are positive, the matrix is positive definite.

7. (d) The Hessian of f is at $(1, 1)$ is

$$\begin{pmatrix} 6 & -3 \\ -3 & 6 \end{pmatrix}$$

Its eigenvalues are $\lambda_1 = 9$, $\lambda_2 = 3$. Since both are positive, the matrix is positive definite and hence $(1, 1)$ is a local minimum.

(e) The Hessian of f at $(1,\ 0,\ 0)$ is

$$\begin{pmatrix} 6 & 0 & 0 \\ 0 & 2 & 1 \\ 0 & 1 & 0 \end{pmatrix}$$

Its eigenvalues are $\lambda_1 = 6$, $\lambda_2 = 1 + \sqrt{2}$, $\lambda_3 = 1 - \sqrt{2}$. Since they differ in sign, $(1,\ 0,\ 0)$ is a saddle point.

8. If A is symmetric positive definite, then all of its eigenvalues are positive. It follows that

$$\det(A) = \lambda_1\lambda_2\cdots\lambda_n > 0$$

The converse is not true. For example if I is the 2×2 identity matrix and we set $A = -I$ then $\det(A) = (-1)\cdot(-1) = 1$, however, A is not positive definite.

9. If A is symmetric positive definite, then all of the eigenvalues $\lambda_1, \lambda_2, \ldots, \lambda_n$ of A are positive. Since 0 is not an eigenvalue, A is nonsingular. The eigenvalues of A^{-1} are $1/\lambda_1,\ 1/\lambda_2, \ldots, 1/\lambda_n$. Thus A^{-1} has positive eigenvalues and hence is positive definite.

10. A^TA is positive semidefinite since

$$\mathbf{x}^TA^TA\mathbf{x} = \|A\mathbf{x}\|^2 \ge 0$$

If A is singular then there exists a nonzero vector $\mathbf{x}$ such that

$$A\mathbf{x} = \mathbf{0}$$

It follows that

$$\mathbf{x}^TA^TA\mathbf{x} = \mathbf{x}^TA^T\mathbf{0} = 0$$

and hence A^TA is not positive definite.

11. Let X be an orthogonal diagonalizing matrix for A. If $\mathbf{x}_1, \ldots, \mathbf{x}_n$ are the column vectors of X then by the remarks following Corollary 6.4.5 we can write

$$A\mathbf{x} = \lambda_1(\mathbf{x}^T\mathbf{x}_1)\mathbf{x}_1 + \lambda_2(\mathbf{x}^T\mathbf{x}_2)\mathbf{x}_2 + \cdots + \lambda_n(\mathbf{x}^T\mathbf{x}_n)\mathbf{x}_n$$

Thus

$$\mathbf{x}^TA\mathbf{x} = \lambda_1(\mathbf{x}^T\mathbf{x}_1)^2 + \lambda_2(\mathbf{x}^T\mathbf{x}_2)^2 + \cdots + \lambda_n(\mathbf{x}^T\mathbf{x}_n)^2$$

12. If A is positive definite, then

$$\mathbf{e}_i^TA\mathbf{e}_i > 0 \quad \text{for} \quad i = 1, \ldots, n$$

but

$$\mathbf{e}_i^TA\mathbf{e}_i = \mathbf{e}_i^T\mathbf{a}_i = a_{ii}$$

13. Let $\mathbf{x}$ be any nonzero vector in R^n and let $\mathbf{y} = S\mathbf{x}$. Since S is nonsingular, $\mathbf{y}$ is nonzero and

$$\mathbf{x}^TS^TAS\mathbf{x} = \mathbf{y}^TA\mathbf{y} > 0$$

Therefore S^TAS is positive definite.

14. If A is symmetric, then by Corollary 6.4.5 there is an orthogonal matrix U that diagonalizes A.

$$A = UDU^T$$

Since A is positive definite, the diagonal elements of D are all positive. If we set

$$Q = UD^{1/2}$$

then the columns of Q are mutually orthogonal and

$$\begin{aligned} A &= (UD^{1/2})((D^{1/2})^TU^T) \\ &= QQ^T \end{aligned}$$

SECTION 7

3. (a)

$$A = \begin{pmatrix} 1 & 0 & 0 & 0 \\ -\frac{1}{2} & 1 & 0 & 0 \\ 0 & -\frac{2}{3} & 1 & 0 \\ 0 & 0 & -\frac{3}{4} & 1 \end{pmatrix} \begin{pmatrix} 2 & -1 & 0 & 0 \\ 0 & \frac{3}{2} & -1 & 0 \\ 0 & 0 & \frac{4}{3} & -1 \\ 0 & 0 & 0 & \frac{5}{4} \end{pmatrix}$$

(b) Since the diagonal entries of U are all positive it follows that A can be reduced to upper triangular form using only row operation III and the pivot elements are all positive. Therefore A must be positive definite.

6. A is symmetric positive definite

$$\langle \mathbf{x}, \mathbf{y} \rangle = \mathbf{x}^T A \mathbf{y}$$

(i) $\langle \mathbf{x}, \mathbf{x} \rangle = \mathbf{x}^T A \mathbf{x} > 0 \quad (\mathbf{x} \neq \mathbf{0})$
since A is positive definite.

(ii) $\langle \mathbf{x}, \mathbf{y} \rangle = \mathbf{x}^T A \mathbf{y} = \mathbf{x}^T A^T \mathbf{y} = (A\mathbf{x})^T \mathbf{y} = \mathbf{y}^T A \mathbf{x} = \langle \mathbf{y}, \mathbf{x} \rangle$

(iii)
$$\begin{aligned} \langle \alpha\mathbf{x} + \beta\mathbf{y}, \mathbf{z} \rangle &= (\alpha\mathbf{x} + \beta\mathbf{y})^T A \mathbf{z} \\ &= \alpha \mathbf{x}^T A \mathbf{z} + \beta \mathbf{y}^T A \mathbf{z} \\ &= \alpha \langle \mathbf{x}, \mathbf{z} \rangle + \beta \langle \mathbf{y}, \mathbf{z} \rangle \end{aligned}$$

7. If $L_1 D_1 U_1 = L_2 D_2 U_2$, then

$$D_2^{-1} L_2^{-1} L_1 D_1 = U_2 U_1^{-1}$$

The left hand side represents a lower triangular matrix and the right hand side represents an upper triangular matrix. Therefore both matrices must be diagonal. Since the matrix U_1 can be transformed into the identity matrix using only row operation III it follows that the diagonal entries of U_1^{-1} must all be 1. Thus

$$U_2 U_1^{-1} = I$$

and hence

$$L_2^{-1} L_1 = D_2 D_1^{-1}$$

Therefore $L_2^{-1} L_1$ is a diagonal matrix and since its diagonal entries must also be 1's we have

$$U_2 U_1^{-1} = I = L_2^{-1} L_1 = D_2 D_1^{-1}$$

or equivalently

$$U_1 = U_2, \qquad L_1 = L_2, \qquad D_1 = D_2$$

8. If A is a positive definite symmetric matrix then A can be factored into a product $A = QDQ^T$ where Q is orthogonal and D is a diagonal matrix whose diagonal elements are all positive. Let E be a diagonal matrix with $e_{ii} = \sqrt{d_{ii}}$ for $i = 1, \ldots, n$. Since $E^T E = E^2 = D$ it follows that

$$A = QE^T EQ^T = (EQ^T)^T (EQ^T) = B^T B$$

where $B = EQ^T$.

9. If B is an $m \times n$ matrix of rank n and $\mathbf{x} \neq \mathbf{0}$, then $B\mathbf{x} \neq \mathbf{0}$. It follows that

$$\mathbf{x}^T B^T B\mathbf{x} = \|B\mathbf{x}\|^2 > 0$$

Therefore $B^T B$ is positive definite.

10. If A is symmetric, then its eigenvalues $\lambda_1, \lambda_2, \ldots, \lambda_n$ are all real and there is an orthogonal matrix Q that diagonalizes A. It follows that

$$A = QDQ^T \qquad \text{and} \qquad e^A = Qe^D Q^T$$

The matrix e^A is symmetric since

$$(e^A)^T = Q(e^D)^T Q^T = Qe^D Q^T = e^A$$

The eigenvalues of e^A are the diagonal entries of e^D

$$\mu_1 = e^{\lambda_1}, \mu_2 = e^{\lambda_2}, \ldots, \mu_n = e^{\lambda_n}$$

Since e^A is symmetric and its eigenvalues are all positive, it follows that e^A is positive definite.

11. Since B is symmetric

$$B^2 = B^T B$$

Since B is also nonsingular, it follows from Theorem 6.7.1 that B^2 is positive definite.

12. (a) A is positive definite since A is symmetric and its eigenvalues $\lambda_1 = \frac{1}{2}$, $\lambda_2 = \frac{3}{2}$ are both positive. If $\mathbf{x} \in R^2$, then

$$\mathbf{x}^T A\mathbf{x} = x_1^2 - x_1 x_2 + x_2^2 = \mathbf{x}^T B\mathbf{x}$$

(b) If $\mathbf{x} \neq \mathbf{0}$, then

$$\mathbf{x}^T B\mathbf{x} = \mathbf{x}^T A\mathbf{x} > 0$$

since A is positive definite. Therefore B is also positive definite. However,

$$B^2 = \begin{pmatrix} 1 & -2 \\ 0 & 1 \end{pmatrix}$$

is not positive definite. Indeed if $\mathbf{x} = (1, 1)^T$, then

$$\mathbf{x}^T B^2 \mathbf{x} = 0$$

13. (a) If A is an symmetric negative definite matrix, then its eigenvalues are all negative. Since the determinant of A is the product of the eigenvalues, it follows that $\det(A)$ will be positive if n is even and negative if n is odd.

(b) Let A_k denote the leading principal submatrix of A of order k and let $\mathbf{x}_1$ be a nonzero vector in R^k. If we set

$$\mathbf{x} = \begin{pmatrix} \mathbf{x}_1 \\ \mathbf{0} \end{pmatrix} \qquad \mathbf{x} \in R^n$$

then

$$\mathbf{x}_1^T A_k \mathbf{x}_1 = \mathbf{x}^T A\mathbf{x} < 0$$

Therefore the leading principal submatrices are all negative definite.

(c) The result in part (c) follows as an immediate consequence of the results from parts (a) and (b).

14. (a) Since $L_{k+1}L_{k+1}^T = A_{k+1}$, we have

$$\begin{pmatrix} L_k & \mathbf{0} \\ \mathbf{x}_k^T & \alpha_k \end{pmatrix}\begin{pmatrix} L_k^T & \mathbf{x}_k \\ \mathbf{0}^T & \alpha_k \end{pmatrix} = \begin{pmatrix} A_k & \mathbf{y}_k \\ \mathbf{y}_k^T & \beta_k \end{pmatrix}$$

$$\begin{pmatrix} L_kL_k^T & L_k\mathbf{x}_k \\ \mathbf{x}_k^TL_k^T & \mathbf{x}_k^T\mathbf{x}_k + \alpha_k^2 \end{pmatrix} = \begin{pmatrix} A_k & \mathbf{y}_k \\ \mathbf{y}_k^T & \beta_k \end{pmatrix}$$

Thus

$$L_k\mathbf{x}_k = \mathbf{y}_k$$

and hence

$$\mathbf{x}_k = L_k^{-1}\mathbf{y}_k$$

Once $\mathbf{x}_k$ has been computed one can solve for α_k.

$$\begin{aligned} \mathbf{x}_k^T\mathbf{x}_k + \alpha_k^2 &= \beta_k \\ \alpha_k &= (\beta_k - \mathbf{x}_k^T\mathbf{x}_k)^{1/2} \end{aligned}$$

(b) **Cholesky Factorization Algorithm**
Set $L_1 = (\sqrt{a_{11}})$
For $k = 1, \ldots, n-1$
(1) Let $\mathbf{y}_k$ be the vector consisting of the first k entries of $\mathbf{a}_{k+1}$ and let β_k be the $(k+1)$st entry of $\mathbf{a}_{k+1}$.
(2) Solve the lower triangular system $L_k\mathbf{x}_k = \mathbf{y}_k$ for $\mathbf{x}_k$.
(3) Set $\alpha_k = (\beta_k - \mathbf{x}_k^T\mathbf{x}_k)^{1/2}$
(4) Set

$$L_{k+1} = \begin{pmatrix} L_k & 0 \\ \mathbf{x}_k^T & \alpha_k \end{pmatrix}$$

End (For Loop)
$L = L_n$

The Cholesky decomposition of A is LL^T.

SECTION 8

7. (b)

$$P = \begin{pmatrix} 1 & 0 & 0 & 0 \\ 0 & 0 & 0 & 1 \\ 0 & 0 & 1 & 0 \\ 0 & 1 & 0 & 0 \end{pmatrix}$$

(c)

$$P = \begin{pmatrix} 1 & 0 & 0 & 0 & 0 \\ 0 & 0 & 1 & 0 & 0 \\ 0 & 1 & 0 & 0 & 0 \\ 0 & 0 & 0 & 1 & 0 \\ 0 & 0 & 0 & 0 & 1 \end{pmatrix}$$

8. It follows from Theorem 6.8.2 that the other two eigenvalues must be

$$\lambda_2 = 2\exp\left(\frac{2\pi i}{3}\right) = -1 + i\sqrt{3}$$

and

$$\lambda_3 = 2\exp\left(\frac{4\pi i}{3}\right) = -1 - i\sqrt{3}$$

9. (a) $A\hat{\mathbf{x}} = \begin{pmatrix} B & O \\ O & C \end{pmatrix} \begin{pmatrix} \mathbf{x} \\ \mathbf{0} \end{pmatrix} = \begin{pmatrix} B\mathbf{x} \\ \mathbf{0} \end{pmatrix} = \begin{pmatrix} \lambda\mathbf{x} \\ \mathbf{0} \end{pmatrix} = \lambda\hat{\mathbf{x}}$

(b) Since B is a positive matrix it has a positive eigenvalue r_1 satisfying the three conditions in Perron's Theorem. Similarly C has a positive eigenvalue r_2 satisfying the conditions of Perron's Theorem. Let $r = \max(r_1, r_2)$. By part (a), r is an eigenvalue of A and condition (iii) of Perron's Theorem implies its multiplicity can be at most 2. (It would have multiplicity 2 in the case that $r_1 = r_2$.) If r_1 has a positive eigenvector $\mathbf{x}$ and r_2 has a positive eigenvector $\mathbf{y}$ then r will have an eigenvector that is either of the form

$$\begin{pmatrix} \mathbf{x} \\ \mathbf{0} \end{pmatrix} \quad \text{or of the form} \quad \begin{pmatrix} \mathbf{0} \\ \mathbf{y} \end{pmatrix}$$

(c) The eigenvalues of A are the eigenvalues of B and C. If $B = C$, then

$$r = r_1 = r_2 \quad \text{(from part (b))}$$

is an eigenvalue of multiplicity 2. If $\mathbf{x}$ is a positive eigenvector of B belonging to r then let

$$\mathbf{z} = \begin{pmatrix} \mathbf{x} \\ \mathbf{x} \end{pmatrix}$$

It follows that

$$A\mathbf{z} = \begin{pmatrix} B & O \\ O & B \end{pmatrix} \begin{pmatrix} \mathbf{x} \\ \mathbf{x} \end{pmatrix} = \begin{pmatrix} B\mathbf{x} \\ B\mathbf{x} \end{pmatrix} = \begin{pmatrix} r\mathbf{x} \\ r\mathbf{x} \end{pmatrix} = r\mathbf{z}$$

Thus $\mathbf{z}$ is a positive eigenvector belonging to r.

10. There are only two possible partitions of the index set $\{1,\ 2\}$. If $I_1 = \{1\}$ and $I_2 = \{2\}$ then A will be reducible provided $a_{12} = 0$. If $I_1 = \{2\}$ and $I_2 = \{1\}$ then A will be reducible provided $a_{21} = 0$. Thus A is reducible if and only if $a_{12}a_{21} = 0$.

11. If A is an irreducible nonnegative 2×2 matrix then it follows from Exercise 10 that $a_{12}a_{21} > 0$. The characteristic polynomial of A

$$p(\lambda) = \lambda^2 - (a_{11} + a_{12})\lambda + (a_{11}a_{22} - a_{12}a_{21})$$

has roots

$$\frac{(a_{11}+a_{22}) \pm \sqrt{(a_{11}+a_{22})^2 - 4(a_{11}a_{22}-a_{12}a_{21})}}{2}$$

The discriminant can be simplified to

$$(a_{11}-a_{22})^2 + 4a_{12}a_{21}.$$

Thus both roots are real. The larger root r_1 is obtained using the + sign.

$$\begin{aligned} r_1 &= \frac{(a_{11}+a_{22}) + \sqrt{(a_{11}-a_{22})^2 + 4a_{12}a_{21}}}{2} \\ &> \frac{a_{11}+a_{22}+|a_{11}-a_{22}|}{2} \\ &= \max(a_{11}, a_{22}) \\ &\geq 0 \end{aligned}$$

Finally r_1 has a positive eigenvector

$$\mathbf{x} = \begin{pmatrix} a_{12} \\ r_1 - a_{11} \end{pmatrix}$$

The case where A has two eigenvalues of equal modulus can only occur when

$$a_{11} = a_{22} = 0$$

In this case $\lambda_1 = \sqrt{a_{21}a_{12}}$ and $\lambda_2 = -\sqrt{a_{21}a_{12}}$.

12. The eigenvalues of A^k are $\lambda_1^k = 1, \lambda_2^k, \ldots, \lambda_n^k$. Clearly $|\lambda_j^k| \leq 1$ for $j = 2, \ldots, n$. However, A^k is a positive matrix and therefore by Perron's theorem $\lambda = 1$ is the dominant eigenvalue and it is a simple root of the characteristic equation for A^k. Therefore $|\lambda_j^k| < 1$ for $j = 2, \ldots, n$ and hence $|\lambda_j| < 1$ for $j = 2, \ldots, n$.

13. (a) It follows from Exercise 12 that $\lambda_1 = 1$ is the dominant eigenvector of A. By Perron's theorem it has a positive eigenvector $\mathbf{x}_1$.

(b) Each $\mathbf{y}_j$ in the chain is a probability vector and hence the coordinates of each vector are nonnegative numbers adding up to 1. Therefore

$$\|\mathbf{y}_j\|_1 = 1 \quad j = 1, 2, \ldots$$

(c) If

$$\mathbf{y}_0 = c_1\mathbf{x}_1 + c_2\mathbf{x}_2 + \cdots + c_n\mathbf{x}_n$$

then

$$\mathbf{y}_k = c_1\mathbf{x}_1 + c_2\lambda_2^k\mathbf{x}_2 + \cdots + c_n\lambda_n^k\mathbf{x}_n$$

and since $\|\mathbf{y}_k\| = 1$ for each k and

$$c_2\lambda_2^k\mathbf{x}_2 + \cdots + c_n\lambda_n^k\mathbf{x}_n \to 0 \quad k \to \infty$$

it follow that $c_1 \neq 0$.

(d) Since

$$\mathbf{y}_k = c_1\mathbf{x}_1 + c_2\lambda_2^k\mathbf{x}_2 + \cdots + c_n\lambda_n^k\mathbf{x}_n$$

and $|\lambda_j| < 1$ for $j = 2, \ldots, n$ it follows that

$$\lim_{k\to\infty} \mathbf{y}_k = c_1\mathbf{x}_1$$

$c_1\mathbf{x}_1$ is the steady-state vector.

(e) Each $\mathbf{y}_k$ is a probability vector and hence the limit vector $c_1\mathbf{x}_1$ must also be a probability vector. Since $\mathbf{x}_1$ is positive it follows that $c_1 > 0$. Thus we have

$$\|c_1\mathbf{x}_1\|_\infty = 1$$

and hence

$$c_1 = \frac{1}{\|\mathbf{x}_1\|_\infty}$$

14. In general if the matrix is nonnegative then there is no guarantee that it has a dominant eigenvalue with a positive eigenvector. So the results from parts (c) and (d) of Exercise 13 would not hold in this case. On the other hand if A^k is a positive matrix for some k, then by Exercise 12, $\lambda_1 = 1$ is the dominant eigenvalue of A and it has a positive eigenvector $\mathbf{x}_1$. Therefore the results from Exercise 13 will be valid in this case.

MATLAB EXERCISES

1. Initially $\mathbf{x} = \mathbf{e}_1$, the standard basis vector, and

$$A\mathbf{x} = \frac{5}{4}\mathbf{e}_1 = \frac{5}{4}\mathbf{x}$$

is in the same direction as $\mathbf{x}$. So $\mathbf{x}_1 = \mathbf{e}_1$ is an eigenvector of A belonging to the eigenvalue $\lambda_1 = \frac{5}{4}$. When the initial vector is rotated so that $\mathbf{x} = \mathbf{e}_2$ the image will be

$$A\mathbf{x} = \frac{3}{4}\mathbf{e}_2 = \frac{3}{4}\mathbf{x}$$

so $\mathbf{x}_2 = \mathbf{e}_2$ is an eigenvector of A belonging to the eigenvalue $\lambda_2 = \frac{3}{4}$. The second diagonal matrix has the same first eigenvalue-eigenvector pair and the second eigenvector is again $\mathbf{x}_2 = \mathbf{e}_2$, however, this time the eigenvalue is negative since $\mathbf{x}_2$ and $A\mathbf{x}_2$ are in opposite directions. In general for any 2×2 diagonal matrix D, the eigenvalues will be d_{11} and d_{22} and the corresponding eigenvectors will be $\mathbf{e}_1$ and $\mathbf{e}_2$.

2. For the identity matrix the eigenvalues are the diagonal entries so $\lambda_1 = \lambda_2 = 1$. In this case not only are $\mathbf{e}_1$ and $\mathbf{e}_2$ eigenvectors, but any vector $\mathbf{x} = x_1\mathbf{e}_1 + x_2\mathbf{e}_2$ is an eigenvector.

3. In this case $\mathbf{x}$ and $A\mathbf{x}$ are equal when $\mathbf{x}$ makes an angle of 45° with the x axis. So $\lambda_1 = 1$ is an eigenvalue with eigenvector

$$\mathbf{x}_1 = \left(\cos\frac{\pi}{4}, \sin\frac{\pi}{4}\right)^T = \left(\frac{1}{\sqrt{2}}, \frac{1}{\sqrt{2}}\right)^T$$

The vectors $\mathbf{x}$ and $A\mathbf{x}$ are unit vectors in opposite directions when $\mathbf{x}$ makes an angle of 135° with the x axis. So $\lambda_2 = -1$ is an eigenvalue and the corresponding eigenvector is

$$\mathbf{x}_2 = \left(\cos\frac{3\pi}{4}, \sin\frac{3\pi}{4}\right)^T = \left(-\frac{1}{\sqrt{2}}, \frac{1}{\sqrt{2}}\right)^T$$

4. In this case $\mathbf{x}$ and $A\mathbf{x}$ are never parallel so A cannot have any real eigenvalues. Therefore the two eigenvalues of A must be complex numbers.

6. For the ninth matrix the vectors $\mathbf{x}$ and $A\mathbf{x}$ are never parallel so A must have complex eigenvalues.

7. The tenth matrix is singular, so one of its eigenvalues is 0. To find the eigenvector using the `eigshow` utility you most rotate $\mathbf{x}$ until $A\mathbf{x}$ coincides with the zero vector. The other eigenvalue of this matrix is $\lambda_2 = 1.5$. Since the eigenvalues are distinct their corresponding eigenvectors must be linearly independent. The next two matrices both have multiple eigenvalues and both are defective. Thus for either matrix any pair of eigenvectors would be linearly dependent.

8. The characteristic polynomial of a 2×2 matrix is a quadratic polynomial and its graph will be a parabola. The eigenvalues will be equal when the graph of the parabola corresponding to the characteristic polynomial has its vertex on the x axis. For a random 2×2 matrix the probability that this will happen should be 0.

11. (a) $A - I$ is a rank one matrix. Therefore the dimension of the eigenspace corresponding to $\lambda = 1$ is 9, the nullity of $A - I$. Thus $\lambda = 1$ has multiplicity at least 9. Since the trace is 20, the remaining eigenvalue $\lambda_{10} = 11$. For symmetric matrices, eigenvalue computations should be quite accurate. Thus one would expect to get nearly full machine accuracy in the computed eigenvalues of A.

(b) The roots of a tenth degree polynomial are quite sensitive, i.e., any small roundoff errors in either the data or in the computations are liable to lead to significant errors in the computed roots. In particular if $p(\lambda)$ has multiple roots, the computed eigenvalues are liable to be complex.

12. (a) When $t = 4$, the eigenvalues change from real to complex. The matrix C corresponding to $t = 4$ has eigenvalues $\lambda_1 = \lambda_2 = 2$. The matrix X of eigenvectors is singular. Thus C does not have two linearly independent eigenvectors and hence must be defective.

(b) The eigenvalues of A correspond to the two points where the graph crosses the x-axis. For each t the graph of the characteristic polynomial will be a parabola. The vertices of these parabolas rise as t increases. When $t = 4$ the vertex will be tangent to the x-axis at $x = 2$. This corresponds to a double eigenvalue. When $t > 4$ the vertex will be above the x-axis. In this case there are no real roots and hence the eigenvalues must be complex.

13. If the rank of B is 2, then its nullity is $4 - 2 = 2$. Thus 0 is an eigenvalue of B and its eigenspace has dimension 2.

14. The reduced row echelon form of C has three lead 1's. Therefore the rank of C is 3 and its nullity is 1. Since $C^4 = O$, all of the eigenvalues of C must be 0. Thus $\lambda = 0$ is an eigenvalue of multiplicity 4 and its eigenspace only has dimension 1. Hence C is defective.

15. In theory A and B should have the same eigenvalues. However for a defective matrix it is difficult to compute the eigenvalues accurately. Thus even though B would be defective if computed in exact arithmetic, the matrix computed using floating point arithmetic may have distinct eigenvalues and the computed matrix X of eigenvectors may turn out to be nonsingular. If, however, `rcond` is very small, this would indicate that the column vectors of X are nearly dependent and hence that B may be defective.

16. (a) Both $A - I$ and $A + I$ have rank 3, so the eigenspaces corresponding to $\lambda_1 = 1$ and $\lambda_2 = -1$ should both have dimension 1.

(b) Since $\lambda_1 + \lambda_2 = 0$ and the sum of all four eigenvalues is 0, it follows that

$$\lambda_3 + \lambda_4 = 0$$

Since $\lambda_1\lambda_2 = -1$ and the product of all four eigenvalues is 1, it follows that

$$\lambda_3\lambda_4 = -1$$

Solving these two equations, we get $\lambda_3 = 1$ and $\lambda_4 = -1$. Thus 1 and -1 are both double eigenvalues. Since their eigenspaces each have dimension 1, the matrix A must be defective.

(d) The computed eigenvectors are linearly independent, but the computed matrix of eigenvectors does not diagonalize A.

17. Since

$$x(2)^2 = \frac{9}{10{,}000}$$

it follows that $x(2) = 0.03$. This proportion should remain constant in future generations. The proportion of genes for color-blindness in the male population should approach 0.03 as the number of generations increases. Thus in the long run 3% of the male population should be color-blind. Since $x(2)^2 = 0.0009$, one would expect that 0.09% of the female population will be color-blind in future generations.

18. (a) By construction S has integer entries and $\det(S) = 1$. It follows that $S^{-1} = \operatorname{adj} S$ will also have integer entries.

19. (a) By construction the matrix A is Hermitian. Therefore its eigenvalues should be real and the matrix X of eigenvectors should be unitary.

(b) The matrix B should be normal. Thus in exact arithmetic $B^H B$ and BB^H should be equal.

20. (a) If $A = USV^T$ then

$$AV = USV^TV = US$$

(b)

$$AV = (A\mathbf{v}_1, A\mathbf{v}_2) \qquad \text{and} \qquad US = (s_1\mathbf{u}_1, s_2\mathbf{u}_2)$$

Since $AV = US$ their corresponding column vectors must be equal. Thus we have

$$A\mathbf{v}_1 = s_1\mathbf{u}_1 \quad \text{and} \quad A\mathbf{v}_2 = s_2\mathbf{u}_2$$

(c) V and U are orthogonal matrices so $\mathbf{v}_1$, $\mathbf{v}_2$ are orthonormal vectors in R^n and $\mathbf{u}_1$, $\mathbf{u}_2$ are orthonormal vectors in R^m. The images $A\mathbf{v}_1$ and $A\mathbf{v}_2$ are orthogonal since

$$(A\mathbf{v}_1)^T A\mathbf{v}_2 = s_1 s_2 \mathbf{u}_1^T \mathbf{u}_2 = 0$$

(d) $\|A\mathbf{v}_1\| = \|s_1\mathbf{u}_1\| = s_1$ and $\|A\mathbf{v}_2\| = \|s_2\mathbf{u}_2\| = s_2$

21. If s_1, s_2 are the singular values of A, $\mathbf{v}_1$, $\mathbf{v}_2$ are the right singular vectors and $\mathbf{u}_1$, $\mathbf{u}_2$, are the corresponding left singular vectors, then the vectors $A\mathbf{x}$ and $A\mathbf{y}$ will be orthogonal when $\mathbf{x} = \mathbf{v}_1$ and $\mathbf{y} = \mathbf{v}_2$. When this happens

$$A\mathbf{x} = A\mathbf{v}_1 = s_1\mathbf{u}_1 \quad \text{and} \quad A\mathbf{y} = A\mathbf{v}_2 = s_2\mathbf{u}_2$$

Thus the image $A\mathbf{x}$ is a vector in the direction of $\mathbf{u}_1$ with length s_1 and the image $A\mathbf{y}$ is a vector in the direction of $\mathbf{u}_2$ with length s_2.
If you rotate the axes a full 360° the image vectors will trace out an ellipse. The major axis of the ellipse will be the line corresponding to the span of $\mathbf{u}_1$ and the diameter of the ellipse along its major axis will be $2s_1$ The minor axis of the ellipse will be the line corresponding to the span of $\mathbf{u}_2$ and the diameter of the ellipse along its minor axis will be $2s_2$.

22. The stationary points of the Hessian are $(-\frac{1}{4}, 0)$ and $(-\frac{71}{4}, 4)$. If the stationary values are substituted into the Hessian, then in each case we can compute the eigenvalues using the MATLAB's *eig* command. If we use the *double* command to view the eigenvalues in numeric format, the displayed values should be 7.6041 and −2.1041 for the first stationary point and −7.6041, 2.1041 for the second stationary point. Thus both stationary points are saddle points.

23. (a) The matrix C is symmetric and hence cannot be defective. The matrix X of eigenvectors should be an orthogonal matrix. The rank of $C - 7I$ is 1 and hence its nullity is 5. Therefore the dimension of the eigenspace corresponding to $\lambda = 7$ is 5.

(b) The matrix C is clearly symmetric and all of its eigenvalues are positive. Therefore C must be positive definite.

(c) In theory R and W should be equal. To see how close the computed matrices actually are, use MATLAB to compute the difference $R - W$.

24. In the $k \times k$ case, U and L will both be bidiagonal. All of the superdiagonal entries of U will be -1 and the diagonal entries will be

$$u_{11} = 2,\ u_{22} = \frac{3}{2},\ u_{33} = \frac{4}{3}, \ldots, u_{kk} = \frac{k+1}{k}$$

L will have 1's on the main diagonal and the subdiagonal entries will be

$$l_{21} = -\frac{1}{2},\ l_{32} = -\frac{2}{3},\ l_{43} = -\frac{3}{4}, \ldots, l_{k,k-1} = -\frac{k-1}{k}$$

Since A can be reduced to upper triangular form U using only row operation III and the diagonal entries of U are all positive, it follows that A must be positive definite.

25. (a) If you subtract 1 from the (6,6) entry of P, the resulting matrix will be singular.

(c) The matrix P is symmetric. The leading principal submatrices of P are all Pascal matrices. If all have determinant equal to 1, then all have positive determinants. Therefore P should be positive definite. The Cholesky factor R is a unit upper triangular matrix. Therefore

$$\det(P) = \det(R^T)\det(R) = 1$$

(d) If one sets $r_{88} = 0$, then R becomes singular. It follows that Q must also be singular since

$$\det(Q) = \det(R^T)\det(R) = 0$$

Since R is upper triangular, when one sets $r_{88} = 0$ it will only affect the $(8, 8)$ entry of the product R^TR. Since R has 1's on the diagonal, changing r_{88} from 1 to 0 will have the effect of decreasing the $(8, 8)$ entry of R^TR by 1.

CHAPTER TEST A

1. The statement is true. If A were singular then we would have

$$\det(A - 0I) = \det(A) = 0$$

so $\lambda = 0$ would have to be an eigenvalue. Therefore if all of the eigenvalues are nonzero, then A cannot be singular.

One could also show that the statement is true by noting that if the eigenvalues of A are all nonzero then

$$\det(A) = \lambda_1\lambda_2\cdots\lambda_n \neq 0$$

and therefore A must be nonsingular.

2. The statement is false in general. A and A^T have the same eigenvalues but generally do not have the same eigenvectors. For example if

$$A = \begin{pmatrix} 1 & 1 \\ 0 & 1 \end{pmatrix} \quad \text{and} \quad \mathbf{e}_1 = \begin{pmatrix} 1 \\ 0 \end{pmatrix}$$

then $A\mathbf{e}_1 = \mathbf{e}_1$ so $\mathbf{e}_1$ is an eigenvector of A. However $\mathbf{e}_1$ is not an eigenvector of A^T since $A^T\mathbf{e}_1$ is not a multiple of $\mathbf{e}_1$.

3. The statement is false in general. The 2×2 identity matrix has eigenvalues $\lambda_1 = \lambda_2 = 1$, but it is not defective.

4. The statement is false. If A is a 4×4 matrix of rank 3, then the nullity of A is 1. Since $\lambda = 0$ is an eigenvalue of multiplicity 3 and the eigenspace has dimension 1, the matrix must be defective.

5. The statement is false. If A is a 4×4 matrix of rank 1, then the nullity of A is 3. Since $\lambda = 0$ is an eigenvalue of multiplicity 3 and the dimension of the eigenspace is also 3, the matrix is diagonalizable.
6. The statement is false in general. The matrix

$$A = \begin{pmatrix} 0 & 1 \\ 0 & 0 \end{pmatrix}$$

has rank 1 even though all of its eigenvalues are 0.
7. The statement is true. If A has singular value decomposition $U\Sigma V^T$, then since U and V are orthogonal matrices, it follows that A and Σ have the same rank. The rank of the diagonal matrix Σ is equal to the number of nonzero singular values.
8. The statement is true. A and T are similar so they have the same eigenvalues. Since T is upper triangular its eigenvalues are its diagonal entries.
9. The statement is true. If A is symmetric positive definite then its eigenvalues are all positive and its determinant is positive. So A must be nonsingular. The inverse of a symmetric matrix is symmetric and the eigenvalues of A^{-1} are the reciprocals of the eigenvalues of A. It follows from Theorem 6.6.2 that A^{-1} must be positive definite.
10. The statement is false in general. For example let

$$A = \begin{pmatrix} -1 & 0 \\ 0 & -1 \end{pmatrix} \quad \text{and} \quad \mathbf{x} = \begin{pmatrix} 1 \\ 1 \end{pmatrix}$$

Although $\det(A) > 0$, the matrix is not positive definite since $\mathbf{x}^T A\mathbf{x} = -2$.

CHAPTER TEST B

1. (a) The eigenvalues of A are $\lambda_1 = 1$, $\lambda_2 = -1$, and $\lambda_3 = 0$,
   (b) Each eigenspace has dimension 1. The vectors that form bases for the eigenspaces are $\mathbf{x}_1 = (1,1,1)^T, \mathbf{x}_2 = (0,1,2)^T, \mathbf{x}_3 = (0,1,1)^T$
   (c)

$$\begin{aligned} A &= XDX^{-1} = \begin{pmatrix} 1 & 0 & 0 \\ 1 & 1 & 1 \\ 1 & 2 & 1 \end{pmatrix} \begin{pmatrix} 1 & 0 & 0 \\ 0 & -1 & 0 \\ 0 & 0 & 0 \end{pmatrix} \begin{pmatrix} 1 & 0 & 0 \\ 0 & -1 & 1 \\ -1 & 2 & -1 \end{pmatrix} \\ A^7 &= XD^7X^{-1} = XDX^{-1} = A \end{aligned}$$

2. Since A has real entries $\lambda_2 = 3 - 2i$ must be an eigenvalue and since A is singular the third eigenvalue is $\lambda_3 = 0$. We can find the last eigenvalue if we make use of the result that the trace of A is equal to the sum of its eigenvalues. Thus we have

$$\text{tr}(A) = 4 = (3 + 2i) + (3 - 2i) + 0 + \lambda_4 = 6 + \lambda_4$$

and hence $\lambda_4 = -2$.

3. **(a)** $\det(A) = \lambda_1\lambda_2\cdots\lambda_n$. If A is nonsingular then $\det(A) \neq 0$ and hence all of the eigenvalues of A must be nonzero.

(b) If λ is an eigenvalue of A then there exists a nonzero vector $\mathbf{x}$ such that $A\mathbf{x} = \lambda\mathbf{x}$. Multiplying both sides of this equation by A^{-1} we get

$$\begin{aligned} A^{-1}A\mathbf{x} &= A^{-1}(\lambda\mathbf{x}) \\ \mathbf{x} &= \lambda A^{-1}\mathbf{x} \\ \frac{1}{\lambda}\mathbf{x} &= A^{-1}\mathbf{x} \end{aligned}$$

and hence $\frac{1}{\lambda}$ is an eigenvalue of A^{-1}.

4. The scalar a is a triple eigenvalue of A. The vector space $N(A-aI)$ consists of all vectors whose third entry is 0. The vectors $\mathbf{e}_1$ and $\mathbf{e}_2$ form a basis for this eigenspace and hence the dimension of the eigenspace is 2. Since the dimension of the eigenspace is less than the multiplicity of the eigenvalue, the matrix must be defective.

5. **(a)**

$$\begin{pmatrix} 4 & 2 & 2 \\ 2 & 10 & 10 \\ 2 & 10 & 14 \end{pmatrix} \to \begin{pmatrix} 4 & 2 & 2 \\ 0 & 9 & 9 \\ 0 & 9 & 13 \end{pmatrix} \to \begin{pmatrix} 4 & 2 & 2 \\ 0 & 9 & 9 \\ 0 & 0 & 4 \end{pmatrix}$$

Since we were able to reduce A to upper triangular form U using only row operation III and the diagonal entries of U are all positive, it follows that A is positive definite.

(b)

$$U = DL^T = \begin{pmatrix} 4 & 0 & 0 \\ 0 & 9 & 0 \\ 0 & 0 & 4 \end{pmatrix} \begin{pmatrix} 1 & \frac{1}{2} & \frac{1}{2} \\ 0 & 1 & 1 \\ 0 & 0 & 1 \end{pmatrix}$$

$$A = LDL^T = \begin{pmatrix} 1 & 0 & 0 \\ \frac{1}{2} & 1 & 0 \\ \frac{1}{2} & 1 & 1 \end{pmatrix} \begin{pmatrix} 4 & 0 & 0 \\ 0 & 9 & 0 \\ 0 & 0 & 4 \end{pmatrix} \begin{pmatrix} 1 & \frac{1}{2} & \frac{1}{2} \\ 0 & 1 & 1 \\ 0 & 0 & 1 \end{pmatrix}$$

(c)

$$L_1 = LD^{\frac{1}{2}} = \begin{pmatrix} 1 & 0 & 0 \\ \frac{1}{2} & 1 & 0 \\ \frac{1}{2} & 1 & 1 \end{pmatrix} \begin{pmatrix} 2 & 0 & 0 \\ 0 & 3 & 0 \\ 0 & 0 & 2 \end{pmatrix} = \begin{pmatrix} 2 & 0 & 0 \\ 1 & 3 & 0 \\ 1 & 3 & 2 \end{pmatrix}$$

$$A = L_1L_1^T = \begin{pmatrix} 2 & 0 & 0 \\ 1 & 3 & 0 \\ 1 & 3 & 2 \end{pmatrix} \begin{pmatrix} 2 & 1 & 1 \\ 0 & 3 & 3 \\ 0 & 0 & 2 \end{pmatrix}$$

6. The first partials of F are

$$f_x = 3x^2y + 2x - 2 \qquad \text{and} \qquad f_y = x^3 + 2y - 1$$

At $(1,0)$ we have $f_x(1,0) = 0$ and $f_y(1,0) = 0$. So $(1,0)$ is a stationary point. The second partials of f are

$$f_{xx} = 6xy + 2, \quad f_{xy} = f_{yx} = 3x^2, \quad f_{yy} = 2$$

At the point $(1,0)$ the Hessian is

$$H = \begin{pmatrix} 2 & 3 \\ 3 & 2 \end{pmatrix}$$

The eigenvalues of H are $\lambda_1 = 5$ and $\lambda_2 = -1$. Since the eigenvalues differ in sign it follows that H is indefinite and hence the stationary point $(1,0)$ is a saddle point.

7. The eigenvalues of A are $\lambda_1 = -1$ and $\lambda_2 = -2$ and the corresponding eigenvectors are $\mathbf{x}_1 = (1,1)^T$ and $\mathbf{x}_2 = (2,3)^T$. The matrix $X = (\mathbf{x}_1, \mathbf{x}_2)$ diagonalizes A and $e^{tA} = Xe^{tD}X^{-1}$. The solution to the initial value problem is

$$\begin{aligned} \mathbf{Y}(t) = e^{tA}\mathbf{Y}_0 &= Xe^{tD}X^{-1}\mathbf{Y}_0 \\ &= \begin{pmatrix} 1 & 2 \\ 1 & 3 \end{pmatrix} \begin{pmatrix} e^{-t} & 0 \\ 0 & e^{-2t} \end{pmatrix} \begin{pmatrix} 3 & -2 \\ -1 & 1 \end{pmatrix} \begin{pmatrix} 1 \\ 2 \end{pmatrix} \\ &= \begin{pmatrix} e^{-t} + 2e^{-2t} \\ e^{-t} + 3e^{-2t} \end{pmatrix} \end{aligned}$$

8. **(a)** Since A is symmetric there is an orthogonal matrix that diagonalizes A. So A cannot be defective and hence the eigenspace corresponding to the triple eigenvalue $\lambda = 0$ (that is, the nullspace of A) must have dimension 3.

(b) Since λ_1 is distinct from the other eigenvalues, the eigenvector $\mathbf{x}_1$ will be orthogonal to $\mathbf{x}_2$, $\mathbf{x}_3$, and $\mathbf{x}_4$.

(c) To construct an orthogonal matrix that diagonalizes A, set $\mathbf{u}_1 = \frac{1}{\|\mathbf{x}_1\|}\mathbf{x}_1$. The vectors $\mathbf{x}_2$, $\mathbf{x}_3$, $\mathbf{x}_4$ form a basis for $N(A)$. Use the Gram-Schmidt process to transform this basis into an orthonormal basis $\{\mathbf{u}_2, \mathbf{u}_3, \mathbf{u}_4\}$. Since the vector $\mathbf{u}_1$ is in $N(A)^\perp$, it follows that $U = (\mathbf{u}_1, \mathbf{u}_2, \mathbf{u}_3, \mathbf{u}_4)$ is an orthogonal matrix and U diagonalizes A.

(d) Since A is symmetric it can be factored into a product $A = QDQ^T$ where Q is orthogonal and D is diagonal. It follows that $e^A = Qe^DQ^T$. The matrix e^A is symmetric since

$$(e^A)^T = Q(e^D)^TQ^T = Qe^DQ^T = e^A$$

The eigenvalues of e^A are $\lambda_1 = e$ and $\lambda_2 = \lambda_3 = \lambda_4 = 1$. Since e^A is symmetric and its eigenvalues are all positive, it follows that e^A is positive definite.

9. **(a)** $\mathbf{u}_1^H\mathbf{z} = 5 - 7i$ and $\mathbf{z}^H\mathbf{u}_1 = 5 + 7i$.
$c_2 = \mathbf{u}_2^H\mathbf{z} == 1 - 5i$.

(b)

$$\begin{aligned} \|\mathbf{z}\|^2 = |c_1|^2 + |c_2|^2 &= (5 - 7i)(5 + 7i) + (1 - 5i)(1 + 5i) \\ &= 25 + 49 + 1 + 25 \end{aligned}$$

$$= 100$$

Therefore $\|\mathbf{z}\| = 10$.

10. (a) The matrix B is symmetric so it eigenvalues are all real. Furthermore, if $\mathbf{x} \neq \mathbf{0}$, then

$$\mathbf{x}^T B\mathbf{x} = \mathbf{x}^T A^T A\mathbf{x} = \|A\mathbf{x}\|^2 \geq 0$$

So B is positive semidefinite and hence its eigenvalues are all nonnegative. Furthermore $N(A)$ has dimension 2, so $\lambda = 0$ is an eigenvalue of multiplicity 2. In summary B is a symmetric positive semidefinite matrix with a double eigenvalue $\lambda = 0$.

(b) The matrix B can be factored into a product QDQ^T where Q is an orthogonal matrix and D is diagonal. It follows that $C = Qe^DQ^T$. So C is symmetric and its eigenvalues are the diagonal entries of e^D which are all positive. Therefore C is a symmetric positive definite matrix.

11. (a) If A has Schur decomposition UTU^H, then U is unitary and T is upper triangular. The matrices A and T are similar so they have the same eigenvalues. Since T is upper triangular it follows that $t_{11}, t_{22}, \ldots, t_{nn}$ are the eigenvalues of both T and A.

(b) If B is Hermitian with Schur decomposition WSW^H, then W is unitary and S is diagonal. The eigenvalues of B are the diagonal entries of S and the column vectors of W are the corresponding eigenvectors.

12. (a) Since A has 3 nonzero singular values, its rank is 3.

(b) If U is the matrix on the left in the given factorization then its first 3 columns, $\mathbf{u}_1$, $\mathbf{u}_2$, $\mathbf{u}_3$ form an orthonormal basis for $R(A)$.

(c) The matrix on the right in the factorization is V^T. The nullity of A is 1 and the vector $\mathbf{v}_4 = (-\frac{1}{2}, \frac{1}{2}, -\frac{1}{2}, \frac{1}{2})^T$ forms a basis for $N(A)$.

(d)

$$B = \sigma_1 \mathbf{u}_1 \mathbf{v}_1^T = 100 \begin{pmatrix} \frac{2}{5} \\ \frac{2}{5} \\ \frac{2}{5} \\ \frac{2}{5} \\ \frac{3}{5} \end{pmatrix} \begin{pmatrix} \frac{1}{2} & \frac{1}{2} & \frac{1}{2} & \frac{1}{2} \end{pmatrix} = \begin{pmatrix} 20 & 20 & 20 & 20 \\ 20 & 20 & 20 & 20 \\ 20 & 20 & 20 & 20 \\ 20 & 20 & 20 & 20 \\ 30 & 30 & 30 & 30 \end{pmatrix}$$

(e) $\|B - A\|_F = \sqrt{10^2 + 10^2} = 10\sqrt{2}$.

CHAPTER 7

SECTION 1

The answers to all of the exercises in this section are included in the text.

SECTION 2

4. (a) (i) $n(mr + mn + n)$ multiplications and $(n-1)m(n+r)$ additions.
(ii) $(mn + nr + mr)$ multiplications and $(n-1)(m+r)$ additions.
(iii) $mn(r+2)$ multiplications and $m(n-1)(r+1)$ additions.

5. (a) The matrix $\mathbf{e}_k\mathbf{e}_i^T$ will have a 1 in the (k,i) position and 0's in all other positions. Thus if $B = I - \alpha\mathbf{e}_k\mathbf{e}_i^T$, then

$$b_{ki} = -\alpha \quad \text{and} \quad b_{sj} = \delta_{sj} \qquad (s,j) \neq (k,i)$$

Therefore $B = E_{ki}$

(b)
$$\begin{aligned} E_{ji}E_{ki} &= (I - \beta\mathbf{e}_j\mathbf{e}_i^T)(I - \alpha\mathbf{e}_k\mathbf{e}_i^T) \\ &= I - \alpha\mathbf{e}_k\mathbf{e}_i^T - \beta\mathbf{e}_j\mathbf{e}_i^T + \alpha\beta\mathbf{e}_j\mathbf{e}_i^T\mathbf{e}_k\mathbf{e}_i^T \\ &= I - (\alpha\mathbf{e}_k + \beta\mathbf{e}_j)\mathbf{e}_i^T \end{aligned}$$

(c)
$$\begin{aligned} (I + \alpha\mathbf{e}_k\mathbf{e}_i^T)E_{ki} &= (I + \alpha\mathbf{e}_k\mathbf{e}_i^T)(I - \alpha\mathbf{e}_k\mathbf{e}_i^T) \\ &= I - \alpha^2\mathbf{e}_k\mathbf{e}_i^T\mathbf{e}_k\mathbf{e}_i^T \\ &= I - \alpha^2(\mathbf{e}_i^T\mathbf{e}_k)\mathbf{e}_k\mathbf{e}_i^T \\ &= I \quad (\text{since } \mathbf{e}_i^T\mathbf{e}_k = 0) \end{aligned}$$

Therefore

$$E_{ki}^{-1} = I + \alpha\mathbf{e}_k\mathbf{e}_i^T$$

6. $\det(A) = \det(L)\det(U) = 1 \cdot \det(U) = u_{11}u_{22}\cdots u_{nn}$

7. Algorithm for solving $LDL^T\mathbf{x} = \mathbf{b}$

For $k = 1, \ldots, n$

Set $y_k = b_k - \sum_{i=1}^{k-1} \ell_{ki} y_i$

Set $z_k = y_k / d_{ii}$

End (For Loop)

For $k = n-1, \ldots, 1$

Set $x_k = z_k - \sum_{j=k+1}^{n} \ell_{jk} x_j$

End (For Loop)

8. (a) Algorithm for solving tridiagonal systems using diagonal pivots

For $k = 1, \ldots, n-1$

Set $m_k := c_k / a_k$

$a_{k+1} := a_{k+1} - m_k b_k$

$d_{k+1} := d_{k+1} - m_k d_k$

End (For Loop)

Set $x_n := d_n / a_n$

For $k = n-1, n-2, \ldots, 1$

Set $x_k := (d_k - b_k x_{k+1}) / a_k$

End (For Loop)

. 9 (b) To solve $A\mathbf{x} = \mathbf{e}_j$, one must first solve $L\mathbf{y} = \mathbf{e}_j$ using forward substitution. From part (a) it follows that this requires $[(n-j)(n-j+1)]/2$ multiplications and $[(n-j-1)(n-j)]/2$ additions. One must then perform back substitution to solve $U\mathbf{x} = \mathbf{y}$. This requires n divisions, $n(n-1)/2$ multiplications and $n(n-1)/2$ additions. Thus altogether, given the LU factorization of A, the number of operations to solve $A\mathbf{x} = \mathbf{e}_j$ is

$$\frac{(n-j)(n-j+1) + n^2 + n}{2} \quad \text{multiplications/divisions}$$

and

$$\frac{(n-j-1)(n-j) + n^2 - n}{2} \quad \text{additions/subtractions}$$

10. Given A^{-1} and $\mathbf{b}$, the multiplication $A^{-1}\mathbf{b}$ requires n^2 scalar multiplications and $n(n-1)$ scalar additions. The same number of operations is required in order to solve $LU\mathbf{x} = \mathbf{b}$ using Algorithm 7.2.2. Thus it is not really worthwhile to calculate A^{-1}, since this calculation requires three times the amount of work it would take to determine L and U.

11. If

$$A(E_1E_2E_3) = L$$

then

$$A = L(E_1E_2E_3)^{-1} = LU$$

The elementary matrices E_1^{-1}, E_2^{-1}, E_3^{-1} will each be upper triangular with ones on the diagonal. Indeed,

$$E_1^{-1} = \begin{pmatrix} 1 & \frac{a_{12}}{a_{11}} & 0 \\ 0 & 1 & 0 \\ 0 & 0 & 1 \end{pmatrix} \quad E_2^{-1} = \begin{pmatrix} 1 & 0 & \frac{a_{13}}{a_{11}} \\ 0 & 1 & 0 \\ 0 & 0 & 1 \end{pmatrix} \quad E_3^{-1} = \begin{pmatrix} 1 & 0 & 0 \\ 0 & 1 & \frac{a_{23}}{a_{22}^{(1)}} \\ 0 & 0 & 1 \end{pmatrix}$$

where $a_{22}^{(1)} = a_{22} - \frac{a_{12}}{a_{11}}$. If we let

$$u_{12} = \frac{a_{12}}{a_{11}}, \quad u_{13} = \frac{a_{13}}{a_{11}}, \quad u_{23} = \frac{a_{23}}{a_{22}^{(1)}}$$

then

$$U = E_3^{-1}E_2^{-1}E_1^{-1} = \begin{pmatrix} 1 & u_{12} & u_{13} \\ 0 & 1 & u_{23} \\ 0 & 0 & 1 \end{pmatrix}$$

SECTION 3

6. (a) $$\left(\begin{array}{ccc|c} 5 & 4 & 7 & 2 \\ 2 & -4 & 3 & -5 \\ 2 & 8 & 6 & 4 \end{array}\right) \rightarrow \left(\begin{array}{ccc|c} 4 & 0 & 4 & 0 \\ 3 & 0 & 6 & -3 \\ 2 & 8 & 6 & 4 \end{array}\right)$$

$$\rightarrow \left(\begin{array}{ccc|c} 2 & 0 & 0 & 2 \\ 3 & 0 & 6 & -3 \\ 2 & 8 & 6 & 4 \end{array}\right)$$

$$\begin{array}{ll} 2x_1 = 2 & x_1 = 1 \\ 3 + 6x_3 = -3 & x_3 = -1 \\ 2 + 8x_2 - 6 = 4 & x_2 = 1 \end{array}$$

$\mathbf{x} = (1,\ 1,\ -1)^T$

(b) The pivot rows were 3, 2, 1 and the pivot columns were 2, 3, 1. Therefore

$$P = \begin{pmatrix} 0 & 0 & 1 \\ 0 & 1 & 0 \\ 1 & 0 & 0 \end{pmatrix} \quad \text{and} \quad Q = \begin{pmatrix} 0 & 0 & 1 \\ 1 & 0 & 0 \\ 0 & 1 & 0 \end{pmatrix}$$

Rearranging the rows and columns of the reduced matrix from part (a), we get

$$U = \begin{pmatrix} 8 & 6 & 2 \\ 0 & 6 & 3 \\ 0 & 0 & 2 \end{pmatrix}$$

The matrix L is formed using the multipliers $-\frac{1}{2}, \frac{1}{2}, \frac{2}{3}$

$$L = \begin{pmatrix} 1 & 0 & 0 \\ -\frac{1}{2} & 1 & 0 \\ \frac{1}{2} & \frac{2}{3} & 1 \end{pmatrix}$$

(c) The system can be solved in 3 steps.

(1) Solve $L\mathbf{y} = P\mathbf{c}$

$$\left(\begin{array}{ccc|c} 1 & 0 & 0 & 2 \\ -\frac{1}{2} & 1 & 0 & -4 \\ \frac{1}{2} & \frac{2}{3} & 1 & 5 \end{array}\right) \qquad \begin{array}{l} y_1 = 2 \\ y_2 = -3 \\ y_3 = 6 \end{array}$$

(2) Solve $U\mathbf{z} = \mathbf{y}$

$$\left(\begin{array}{ccc|c} 8 & 6 & 2 & 2 \\ 0 & 6 & 3 & -3 \\ 0 & 0 & 2 & 6 \end{array}\right) \qquad \begin{array}{l} z_1 = 1 \\ z_2 = -2 \\ z_3 = 3 \end{array}$$

(3) Set $\mathbf{x} = Q\mathbf{z}$

$$\mathbf{x} = \begin{pmatrix} 0 & 0 & 1 \\ 1 & 0 & 0 \\ 0 & 1 & 0 \end{pmatrix} \begin{pmatrix} 1 \\ -2 \\ 3 \end{pmatrix} = \begin{pmatrix} 3 \\ 1 \\ -2 \end{pmatrix}$$

SECTION 4

3. Let $\mathbf{x}$ be a nonzero vector in R^2

$$\frac{\|A\mathbf{x}\|_2}{\|\mathbf{x}\|_2} = \frac{|x_1|}{\sqrt{x_1^2 + x_2^2}} \le 1$$

Therefore

$$\|A\|_2 = \max_{\mathbf{x} \neq \mathbf{0}} \frac{\|A\mathbf{x}\|_2}{\|\mathbf{x}\|_2} \le 1$$

On the other hand

$$\|A\|_2 \ge \frac{\|A\mathbf{e}_1\|_2}{\|\mathbf{e}_1\|_2} = 1$$

Therefore $\|A\|_2 = 1$.

4. (a) D has singular value decomposition $U\Sigma V^T$ where the diagonal entries of Σ are $\sigma_1 = 5$, $\sigma_2 = 4$, $\sigma_3 = 3$, $\sigma_4 = 2$ and

$$U = \begin{pmatrix} 0 & 0 & 1 & 0 \\ 1 & 0 & 0 & 0 \\ 0 & 0 & 0 & 1 \\ 0 & 1 & 0 & 0 \end{pmatrix}, \quad V = \begin{pmatrix} 0 & 0 & 1 & 0 \\ -1 & 0 & 0 & 0 \\ 0 & 0 & 0 & -1 \\ 0 & 1 & 0 & 0 \end{pmatrix}$$

(b) $\|D\|_2 = \sigma_1 = 5$

5. If D is diagonal then its singular values are the square roots of the eigenvalues of $D^TD = D^2$. The eigenvalues of D^2 are $d_{11}^2, d_{22}^2, \ldots, d_{nn}^2$ and hence it follows that

$$\|D\|_2 = \sigma_1 = \max_{1\le i\le n} |d_{ii}|$$

6. It follows from Theorem 7.4.2 that

$$\|D\|_1 = \|D\|_\infty = \max_{1\le i\le n} |d_{ii}|$$

and it follows from Exercise 5 that this is also the value of $\|D\|_2$. Thus for a diagonal matrix all 3 norms are equal.

8. (a) If $\|\cdot\|_M$ and $\|\cdot\|_V$ are compatible, then for any nonzero vector $\mathbf{x}$,

$$\|\mathbf{x}\|_V = \|I\mathbf{x}\|_V \le \|V\|_M\|\mathbf{x}\|_V$$

Dividing by $\|\mathbf{x}\|_V$ we get

$$1 \le \|I\|_M$$

(b) If $\|\cdot\|_M$ is subordinate to $\|\cdot\|_V$, then

$$\frac{\|I\mathbf{x}\|_V}{\|\mathbf{x}\|_V} = 1$$

for all nonzero vectors $\mathbf{x}$ and it follows that

$$\|I\|_M = \max_{\mathbf{x}\ne 0} \frac{\|I\mathbf{x}\|_V}{\|\mathbf{x}\|_V} = 1$$

9. (a) $\|X\|_\infty = \|\mathbf{x}\|_\infty$ since the ith row sum is just $|x_i|$ for each i.

(b) The 1-norm of a matrix is equal to the maximum of the 1-norm of its column vectors. Since X only has one column its 1-norm is equal to the 1-norm of that column vector.

11. Let $\mathbf{x}$ be a nonzero vector in R^n

$$\begin{aligned}
\frac{\|A\mathbf{x}\|_\infty}{\|\mathbf{x}\|_\infty} &= \frac{\max\limits_{1\le i\le m} \left|\sum\limits_{j=1}^n a_{ij}x_j\right|}{\max\limits_{1\le j\le n} |x_j|} \\
&\le \frac{\max\limits_{1\le j\le n} |x_j| \max\limits_{1\le i\le m} \left|\sum\limits_{j=1}^n a_{ij}\right|}{\max\limits_{1\le j\le n} |x_j|} \\
&= \max_{1\le i\le m} \left|\sum_{j=1}^n a_{ij}\right| \\
&\le \max_{1\le i\le m} \sum_{j=1}^n |a_{ij}|
\end{aligned}$$

Therefore

$$\|A\|_\infty = \max_{\mathbf{x}\neq 0} \frac{\|A\mathbf{x}\|}{\|\mathbf{x}\|} \leq \max_{1\leq i\leq m} \sum_{j=1}^{n} |a_{ij}|$$

Let k be the index of the row of A for which $\sum_{j=1}^{n} |a_{ij}|$ is a maximum. Define $x_j = \operatorname{sgn} a_{kj}$ for $j = 1, \ldots, n$ and let $\mathbf{x} = (x_1, \ldots, x_n)^T$. Note that $\|\mathbf{x}\|_\infty = 1$ and $a_{kj}x_j = |a_{kj}|$ for $j = 1, \ldots, n$. Thus

$$\|A\|_\infty \geq \|A\mathbf{x}\|_\infty = \max_{1\leq i\leq n} \left| \sum_{j=1}^{n} a_{ij}x_j \right| \geq \sum_{j=1}^{n} |a_{kj}|$$

Therefore

$$\|A\|_\infty = \max_{1\leq i\leq m} \left(\sum_{j=1}^{n} |a_{ij}| \right)$$

12. $\|A\|_F = \left(\sum_j \sum_i a_{ij}^2 \right)^{1/2} = \left(\sum_i \sum_j a_{ij}^2 \right)^{1/2} = \|A^T\|_F$

13. $\|A\|_\infty = \max_{1\leq i\leq n} \sum_{j=1}^{n} |a_{ij}| = \max_{1\leq i\leq n} \sum_{j=1}^{n} |a_{ji}| = \|A\|_1$

14. $\|A\|_2 = \sigma_1 = 5$ and

$$\|A\|_F = (\sigma_1^2 + \sigma_2^2 + \sigma_3^2 + \sigma_4^2 + \sigma_5^2)^{\frac{1}{2}} = 6$$

15. (a) Let $k = \min(m, n)$.

(6) $$\|A\|_2 = \sigma_1 \leq (\sigma_1^2 + \sigma_2^2 + \cdots + \sigma_k^2)^{\frac{1}{2}} = \|A\|_F$$

(b) Equality will hold in (6) if $\sigma_2^2 = \cdots = \sigma_k^2 = 0$. It follows then that $\|A\|_2 = \|A\|_F$ if and only if the matrix A has rank 1.

16. Since

$$\{\mathbf{x} \mid \|\mathbf{x}\| = 1\} = \{\mathbf{x} \mid \mathbf{x} = \frac{1}{\|\mathbf{y}\|}\mathbf{y}, \quad \mathbf{y} \in R^n \text{ and } \mathbf{y} \neq \mathbf{0}\}$$

$$\begin{aligned} \|A\|_M &= \max_{\mathbf{y}\neq\mathbf{0}} \frac{\|A\mathbf{y}\|}{\|\mathbf{y}\|} \\ &= \max_{\mathbf{y}\neq\mathbf{0}} \left\| A \left(\frac{1}{\|\mathbf{y}\|}\mathbf{y} \right) \right\| \\ &= \max_{\|\mathbf{x}\|=1} \|A\mathbf{x}\| \end{aligned}$$

17. If $\mathbf{x}$ is a unit eigenvector belonging to the eigenvalue λ, then

$$|\lambda| = \|\lambda\mathbf{x}\| = \|A\mathbf{x}\| \leq \|A\|_M \|\mathbf{x}\| = \|A\|_M$$

18. If A is a stochastic matrix then $\|A\|_1 = 1$. It follows from Exercise 17 that if λ is an eigenvalue of A then

$$|\lambda| \le \|A\|_1 = 1$$

19. (b) $\|A\mathbf{x}\|_2 \le n^{1/2}\|A\mathbf{x}\|_\infty \le n^{1/2}\|A\|_\infty\|\mathbf{x}\|_\infty \le n^{1/2}\|A\|_\infty\|\mathbf{x}\|_2$

(c) Let $\mathbf{x}$ be any nonzero vector in R^n. It follows from part (a) that

$$\frac{\|A\mathbf{x}\|_\infty}{\|\mathbf{x}\|_\infty} \le n^{1/2}\|A\|_2$$

and it follows from part (b) that

$$\frac{\|A\mathbf{x}\|_2}{\|\mathbf{x}\|_2} \le n^{1/2}\|A\|_\infty$$

Consequently

$$\|A\|_\infty = \max_{\mathbf{x}\ne\mathbf{0}} \frac{\|A\mathbf{x}\|_\infty}{\|\mathbf{x}\|_\infty} \le n^{1/2}\|A\|_2$$

and

$$\|A\|_2 = \max_{\mathbf{x}\ne\mathbf{0}} \frac{\|A\mathbf{x}\|_2}{\|\mathbf{x}\|_2} \le n^{1/2}\|A\|_\infty$$

Thus

$$n^{-1/2}\|A\|_2 \le \|A\|_\infty \le n^{1/2}\|A\|_2$$

20. Let A be a symmetric matrix with orthonormal eigenvectors $\mathbf{u}_1, \ldots, \mathbf{u}_n$. If $\mathbf{x} \in R^n$ then by Theorem 5.5.2

$$\mathbf{x} = c_1\mathbf{u}_1 + c_2\mathbf{u}_2 + \cdots + c_n\mathbf{u}_n$$

where $c_i = \mathbf{u}_i^T\mathbf{x}$, $i = 1, \ldots, n$.

(a) $A\mathbf{x} = c_1A\mathbf{u}_1 + c_2A\mathbf{u}_2 + \cdots + c_nA\mathbf{u}_n$

$\quad = c_1\lambda_1\mathbf{u}_1 + c_2\lambda_2\mathbf{u}_2 + \cdots + c_n\lambda_n\mathbf{u}_n.$

It follows from Parseval's formula that

$$\|A\mathbf{x}\|_2^2 = \sum_{i=1}^n (\lambda_i c_i)^2$$

(b) It follows from part (a) that

$$\min_{1\le i\le n} |\lambda_i| \left(\sum_{i=1}^n c_i^2\right)^{1/2} \le \|A\mathbf{x}\|_2 \le \max_{1\le i\le n} |\lambda_i| \left(\sum_{i=1}^n c_i^2\right)^{1/2}$$

Using Parseval's formula we see that

$$\left(\sum_{j=1}^n c_i^2\right)^{1/2} = \|\mathbf{x}\|_2$$

and hence for any nonzero vector $\mathbf{x}$ we have

$$\min_{1\le i\le n} |\lambda_i| \le \frac{\|A\mathbf{x}\|_2}{\|\mathbf{x}\|_2} \le \max_{1\le i\le n} |\lambda_i|$$

(c) If

$$|\lambda_k| = \max_{1\le i\le n} |\lambda_i|$$

and $\mathbf{x}_k$ is an eigenvector belonging to λ_k, then

$$\frac{\|A\mathbf{x}_k\|_2}{\|\mathbf{x}_k\|_2} = |\lambda_k| = \max_{1\le i\le n} |\lambda_i|$$

and hence it follows from part (b) that

$$\|A\|_2 = \max_{1\le i\le n} |\lambda_i|$$

21.

$$A^{-1} = \begin{pmatrix} 100 & 99 \\ 100 & 100 \end{pmatrix}$$

22. Let A be the coefficient matrix of the first system and A' be the coefficient matrix of the second system. If $\mathbf{x}$ is the solution to the first system and $\mathbf{x}'$ is the solution to the second system then

$$\frac{\|\mathbf{x}-\mathbf{x}'\|_\infty}{\|\mathbf{x}\|_\infty} \approx 3.03$$

while

$$\frac{\|A-A'\|_\infty}{\|A\|_\infty} \approx 0.014$$

The systems are ill-conditioned in the sense that a relative change of 0.014 in the coefficient matrix results in a relative change of 3.03 in the solution.

24. $\mathrm{cond}(A) = \|A\|_M\|A^{-1}\|_M \ge \|AA^{-1}\|_M = \|I\|_M = 1.$

26. The given conditions allow us to determine the singular values of the matrix. Indeed, $\sigma_1 = \|A\|_2 = 8$ and since

$$\frac{\sigma_1}{\sigma_3} = \mathrm{cond}_2(A) = 2$$

it follows that $\sigma_3 = 4$. Finally

$$\begin{aligned} \sigma_1^2+\sigma_2^2+\sigma_3^2 &= \|A\|_F^2 \\ 64+\sigma_2^2+16 &= 144 \end{aligned}$$

and hence $\sigma_2 = 8$.

27. (c) $\dfrac{1}{\mathrm{cond}_\infty(A)}\dfrac{\|\mathbf{r}\|_\infty}{\|\mathbf{b}\|_\infty} \le \dfrac{\|\mathbf{x}-\mathbf{x}'\|_\infty}{\|\mathbf{x}\|_\infty} \le \mathrm{cond}_\infty(A)\dfrac{\|\mathbf{r}\|_\infty}{\|\mathbf{b}\|_\infty}$

$$0.0006 = \tfrac{1}{20}(0.012) \le \tfrac{\|\mathbf{x}-\mathbf{x}'\|_\infty}{\|\mathbf{x}\|_\infty} \le 20(0.012) = 0.24$$

31. $\mathrm{cond}(AB) = \|AB\|\,\|(AB)^{-1}\| \le \|A\|\,\|B\|\,\|B^{-1}\|\,\|A^{-1}\| = \mathrm{cond}\,(A)\,\mathrm{cond}(B)$

32. It follows from Exercises 5 and 6 that

$$\|D\|_1 = \|D\|_2 = \|D\|_\infty = d_{\max}$$

and

$$\|D^{-1}\|_1 = \|D^{-1}\|_2 = \|D^{-1}\|_\infty = \frac{1}{d_{\min}}$$

Therefore the condition number of D will be $\dfrac{d_{\max}}{d_{\min}}$ no matter which of the 3 norms is used.

33. (a) For any vector $\mathbf{x}$

$$\|Q\mathbf{x}\|_2 = \|\mathbf{x}\|_2$$

Thus if $\mathbf{x}$ is nonzero, then

$$\frac{\|Q\mathbf{x}\|_2}{\|\mathbf{x}\|_2} = 1$$

and hence

$$\|Q\|_2 = \max_{\mathbf{x}\neq\mathbf{0}} \frac{\|Q\mathbf{x}\|_2}{\|\mathbf{x}\|_2} = 1$$

(b) The matrix $Q^{-1} = Q^T$ is also orthogonal and hence by part (a) we have

$$\|Q^{-1}\|_2 = 1$$

Therefore

$$\text{cond}_2(Q) = 1$$

(c) $$\frac{1}{\text{cond}_2(Q)}\frac{\|\mathbf{r}\|_2}{\|\mathbf{b}\|_2} \le \frac{\|\mathbf{e}\|_2}{\|\mathbf{x}\|_2} \le \text{cond}_2(Q)\frac{\|\mathbf{r}\|_2}{\|\mathbf{b}\|_2}$$

Since $\text{cond}_2(Q) = 1$, it follows that

$$\frac{\|\mathbf{e}\|_2}{\|\mathbf{x}\|_2} = \frac{\|\mathbf{r}\|_2}{\|\mathbf{b}\|_2}$$

34. (a) If $\mathbf{x}$ is any vector in R^r, then $A\mathbf{x}$ is a vector in R^n and

$$\|QA\mathbf{x}\|_2 = \|A\mathbf{x}\|_2$$

Thus for any nonzero vector $\mathbf{x}$

$$\frac{\|QA\mathbf{x}\|_2}{\|\mathbf{x}\|_2} = \frac{\|A\mathbf{x}\|_2}{\|\mathbf{x}\|_2}$$

and hence

$$\begin{aligned}\|QA\|_2 &= \max_{\mathbf{x}\neq\mathbf{0}} \frac{\|QA\mathbf{x}\|_2}{\|\mathbf{x}\|_2} \\ &= \max_{\mathbf{x}\neq\mathbf{0}} \frac{\|A\mathbf{x}\|_2}{\|\mathbf{x}\|_2} \\ &= \|A\|_2\end{aligned}$$

(b) For each nonzero vector $\mathbf{x}$ in R^n set $\mathbf{y} = V\mathbf{x}$. Since V is nonsingular it follows that $\mathbf{y}$ is nonzero. Furthermore

$$\{\mathbf{y}\,|\,\mathbf{y} = V\mathbf{x} \text{ and } \mathbf{x}\neq\mathbf{0}\} = \{\mathbf{x}\,|\,\mathbf{x}\neq\mathbf{0}\}$$

since any nonzero $\mathbf{y}$ can be written as

$$\mathbf{y} = V\mathbf{x} \qquad \text{where } \mathbf{x} = V^T\mathbf{y}$$

It follows that if $\mathbf{x} \neq \mathbf{0}$ and $\mathbf{y} = V\mathbf{x}$, then

$$\frac{\|AV\mathbf{x}\|_2}{\|\mathbf{x}\|_2} = \frac{\|AV\mathbf{x}\|_2}{\|V\mathbf{x}\|_2} = \frac{\|A\mathbf{y}\|_2}{\|\mathbf{y}\|_2}$$

and hence

$$\|AV\|_2 = \max_{\mathbf{x}\neq\mathbf{0}} \frac{\|AV\mathbf{x}\|_2}{\|\mathbf{x}\|_2} = \max_{\mathbf{y}\neq\mathbf{0}} \frac{\|A\mathbf{y}\|_2}{\|\mathbf{y}\|_2} = \|A\|_2$$

(c) It follows from parts (a) and (b) that

$$\|QAV\|_2 = \|Q(AV)\|_2 = \|AV\|_2 = \|A\|_2$$

35. (a) If A has singular value decomposition $U\Sigma V^T$, then it follows from the Cauchy-Schwarz inequality that

$$|\mathbf{x}^T A\mathbf{y}| \leq \|\mathbf{x}\|_2\|A\mathbf{y}\|_2 \leq \|\mathbf{x}\|_2\|\mathbf{y}\|_2\|A\|_2 = \sigma_1\|\mathbf{x}\|_2\|\mathbf{y}\|_2$$

Thus if $\mathbf{x}$ and $\mathbf{y}$ are nonzero vectors, then

$$\frac{|\mathbf{x}^T A\mathbf{y}|}{\|\mathbf{x}\|_2\|\mathbf{y}\|_2} \leq \sigma_1$$

(b) If we set $\mathbf{x}_1 = \mathbf{u}_1$ and $\mathbf{y}_1 = \mathbf{v}_1$, then

$$\|\mathbf{x}_1\|_2 = \|\mathbf{u}_1\|_2 = 1 \qquad \text{and} \qquad \|\mathbf{y}_1\|_2 = \|\mathbf{v}_1\|_2 = 1$$

and

$$A\mathbf{y}_1 = A\mathbf{v}_1 = \sigma_1\mathbf{u}_1$$

Thus

$$\mathbf{x}_1^T A\mathbf{y}_1 = \mathbf{u}_1^T(\sigma_1\mathbf{u}_1) = \sigma_1$$

and hence

$$\frac{|\mathbf{x}_1^T A\mathbf{y}_1|}{\|\mathbf{x}_1\|_2\|\mathbf{y}_1\|_2} = \sigma_1$$

Combining this with the result from part (a) we have

$$\max_{\mathbf{x}\neq\mathbf{0},\mathbf{y}\neq\mathbf{0}} \frac{|\mathbf{x}^T A\mathbf{y}|}{\|\mathbf{x}\|_2\|\mathbf{y}\|_2} = \sigma_1$$

36. For each nonzero vector $\mathbf{x}$ in R^n

$$\frac{\|A\mathbf{x}\|_2}{\|\mathbf{x}\|_2} = \frac{\|U\Sigma V^T\mathbf{x}\|_2}{\|\mathbf{x}\|_2} = \frac{\|\Sigma V^T\mathbf{x}\|_2}{\|V^T\mathbf{x}\|_2} = \frac{\|\Sigma\mathbf{y}\|_2}{\|\mathbf{y}\|_2}$$

where $\mathbf{y} = V^T\mathbf{x}$. Thus

$$\min_{\mathbf{x}\neq 0} \frac{\|A\mathbf{x}\|_2}{\|\mathbf{x}\|_2} = \min_{\mathbf{y}\neq 0} \frac{\|\Sigma\mathbf{y}\|_2}{\|\mathbf{y}\|_2}$$

For any nonzero vector $\mathbf{y} \in R^n$

$$\frac{\|\Sigma\mathbf{y}\|_2}{\|\mathbf{y}\|_2} = \frac{\left(\sum_{i=1}^{n} \sigma_i^2 y_i^2\right)^{1/2}}{\left(\sum_{i=1}^{n} y_i^2\right)^{1/2}} \geq \frac{\sigma_n \|\mathbf{y}\|_2}{\|\mathbf{y}\|_2} = \sigma_n$$

Thus

$$\min \frac{\|\Sigma\mathbf{y}\|_2}{\|\mathbf{y}\|_2} \geq \sigma_n$$

On the other hand

$$\min_{\mathbf{y}\neq\mathbf{0}} \frac{\|\Sigma\mathbf{y}\|_2}{\|\mathbf{y}\|_2} \leq \frac{\|\Sigma\mathbf{e}_n\|_2}{\|\mathbf{e}_n\|_2} = \sigma_n$$

Therefore

$$\min_{\mathbf{x}\neq\mathbf{0}} \frac{\|A\mathbf{x}\|_2}{\|\mathbf{x}\|_2} = \min_{\mathbf{y}\neq\mathbf{0}} \frac{\|\Sigma\mathbf{y}\|_2}{\|\mathbf{y}\|_2} = \sigma_n$$

37. For any nonzero vector $\mathbf{x}$

$$\frac{\|A\mathbf{x}\|_2}{\|\mathbf{x}\|_2} \leq \|A\|_2 = \sigma_1$$

It follows from Exercise 33 that

$$\frac{\|A\mathbf{x}\|_2}{\|\mathbf{x}\|_2} \geq \sigma_n$$

Thus if $\mathbf{x} \neq \mathbf{0}$, then

$$\sigma_n\|\mathbf{x}\|_2 \leq \|A\mathbf{x}\|_2 \leq \sigma_1\|\mathbf{x}\|_2$$

Clearly this inequality is also valid if $\mathbf{x} = \mathbf{0}$.

38. (a) It follows from Exercise 34 that

$$\|QA\|_2 = \|A\|_2 \quad \text{and} \quad \|A^{-1}Q^T\|_2 = \|A^{-1}\|_2$$
$$\|AQ\|_2 = \|A\|_2 \quad \text{and} \quad \|Q^TA^{-1}\|_2 = \|A^{-1}\|_2$$

Thus

$$\begin{aligned} \text{cond}_2(QA) &= \|QA\|_2\|A^{-1}Q^T\|_2 = \text{cond}_2(A) \\ \text{cond}_2(AQ) &= \|AQ\|_2\|Q^TA^{-1}\|_2 = \text{cond}_2(A) \end{aligned}$$

(b) It follows from Exercise 34 that

$$\|B\|_2 = \|A\|_2$$

and

$$\|B^{-1}\|_2 = \|Q^TA^{-1}Q\|_2 = \|A^{-1}\|_2$$

Therefore

$$\text{cond}_2(B) = \text{cond}_2(A)$$

39. If A is a symmetric $n \times n$ matrix, then there exists an orthogonal matrix Q that diagonalizes A.

$$Q^T AQ = D$$

The diagonal elements of D are the eigenvalues of A. Since A is symmetric and nonsingular its eigenvalues are all nonzero real numbers. It follows from Exercise 38 that

$$\text{cond}_2(A) = \text{cond}_2(D)$$

and it follows from Exercise 32 that

$$\text{cond}_2(D) = \frac{\lambda_{\max}}{\lambda_{\min}}$$

SECTION 5

7. (b)

$$G = \begin{pmatrix} \frac{1}{\sqrt{2}} & \frac{1}{\sqrt{2}} \\ \frac{1}{\sqrt{2}} & -\frac{1}{\sqrt{2}} \end{pmatrix}$$

$$(GA \mid G\mathbf{b}) = \left(\begin{array}{cc|c} \sqrt{2} & 3\sqrt{2} & 3\sqrt{2} \\ 0 & \sqrt{2} & 2\sqrt{2} \end{array} \right), \quad \mathbf{x} = \begin{pmatrix} -3 \\ 2 \end{pmatrix}$$

(c)

$$G = \begin{pmatrix} \frac{4}{5} & 0 & -\frac{3}{5} \\ 0 & 1 & 0 \\ -\frac{3}{5} & 0 & -\frac{4}{5} \end{pmatrix}$$

$$(GA \mid G\mathbf{b}) = \left(\begin{array}{ccc|c} 5 & -5 & 2 & 1 \\ 0 & 1 & 3 & 2 \\ 0 & 0 & 1 & -2 \end{array} \right), \quad \mathbf{x} = \begin{pmatrix} 9 \\ 8 \\ -2 \end{pmatrix}$$

12. (a)
$$\begin{aligned} \|\mathbf{x} - \mathbf{y}\|^2 &= (\mathbf{x} - \mathbf{y})^T(\mathbf{x} - \mathbf{y}) \\ &= \mathbf{x}^T\mathbf{x} - \mathbf{x}^T\mathbf{y} - \mathbf{y}^T\mathbf{x} + \mathbf{y}^T\mathbf{y} \\ &= 2\mathbf{x}^T\mathbf{x} - 2\mathbf{y}^T\mathbf{x} \\ &= 2(\mathbf{x} - \mathbf{y})^T\mathbf{x} \end{aligned}$$

(b) It follows from part (a) that

$$2\mathbf{u}^T\mathbf{x} = \frac{2}{\|\mathbf{x} - \mathbf{y}\|}(\mathbf{x} - \mathbf{y})^T\mathbf{x} = \|\mathbf{x} - \mathbf{y}\|$$

Thus

$$2\mathbf{u}\mathbf{u}^T\mathbf{x} = (2\mathbf{u}^T\mathbf{x})\mathbf{u} = \mathbf{x} - \mathbf{y}$$

and hence

$$Q\mathbf{x} = (I - 2\mathbf{u}\mathbf{u}^T)\mathbf{x} = \mathbf{x} - (\mathbf{x} - \mathbf{y}) = \mathbf{y}$$

13. (a) $Q\mathbf{u} = (I - 2\mathbf{u}\mathbf{u}^T)\mathbf{u} = \mathbf{u} - 2(\mathbf{u}^T\mathbf{u})\mathbf{u} = -\mathbf{u}$
The eigenvalue is $\lambda = -1$.

(b) $Q\mathbf{z} = (I - 2\mathbf{u}\mathbf{u}^T)\mathbf{z} = \mathbf{z} - 2(\mathbf{u}^T\mathbf{z})\mathbf{u} = \mathbf{z}$
Therefore $\mathbf{z}$ is an eigenvector belonging to the eigenvalue $\lambda = 1$.

(c) The eigenspace corresponding to $\lambda = 1$ is

$$N(Q - I) = N(-2\mathbf{u}\mathbf{u}^T) = N(\mathbf{u}\mathbf{u}^T)$$

The matrix $\mathbf{u}\mathbf{u}^T$ has rank 1 and hence its nullity must be $n - 1$. Thus the dimension of the eigenspace corresponding to $\lambda = 1$ is $n - 1$. Therefore the multiplicity of the eigenvalue must be at least $n - 1$. Since we know that -1 is an eigenvalue, it follows that $\lambda = 1$ must have multiplicity $n - 1$. Since the determinant is equal to the product of the eigenvalues we have

$$\det(Q) = -1 \cdot (1)^n = -1$$

14. If R is a plane rotation then expanding its determine by cofactors we see that

$$\det(R) = \cos^2\theta + \sin^2\theta = 1$$

By Exercise 13(c) an elementary orthogonal matrix has determinant equal to -1, so it follows that a plane rotation cannot be an elementary orthogonal matrix.

15. (a) Let $Q = Q_1^T Q_2 = R_1 R_2^{-1}$. The matrix Q is orthogonal and upper triangular. Since Q is upper triangular, Q^{-1} must also be upper triangular. However

$$Q^{-1} = Q^T = (R_1 R_2^{-1})^T$$

which is lower triangular. Therefore Q must be diagonal.

(b) $R_1 = (Q_1^T Q_2)R_2 = QR_2$. Since

$$|q_{ii}| = \|Q\mathbf{e}_i\| = \|\mathbf{e}_i\| = 1$$

it follows that $q_{ii} = \pm 1$ and hence the ith row of R_1 is ± 1 times the ith row of R_2.

16. Since $\mathbf{x}$ and $\mathbf{y}$ are nonzero vectors, there exist Householder matrices H_1 and H_2 such that

$$H_1\mathbf{x} = \|\mathbf{x}\|\mathbf{e}_1^{(m)} \quad \text{and} \quad H_2\mathbf{y} = \|\mathbf{y}\|\mathbf{e}_2^{(n)}$$

where $\mathbf{e}_1^{(m)}$ and $\mathbf{e}_1^{(n)}$ denote the first column vectors of the $m \times m$ and $n \times n$ identity matrices. It follows that

$$\begin{aligned} H_1 A H_2 &= H_1 \mathbf{x}\mathbf{y}^T H_2 \\ &= (H_1\mathbf{x})(H_2\mathbf{y})^T \\ &= \|\mathbf{x}\|\,\|\mathbf{y}\|\mathbf{e}_1^{(m)}(\mathbf{e}_1^{(n)})^T \end{aligned}$$

Set

$$\Sigma = \|\mathbf{x}\|\,\|\mathbf{y}\|\mathbf{e}_1^{(m)}(\mathbf{e}_1^{(n)})^T$$

Σ is an $m \times n$ matrix whose entries are all zero except for the $(1, 1)$ entry which equals $\|\mathbf{x}\|\,\|\mathbf{y}\|$. We have then

$$H_1 A H_2 = \Sigma$$

Since H_1 and H_2 are both orthogonal and symmetric it follows that A has singular value decomposition $H_1 \Sigma H_2$.

17. In constructing the Householder matrix we set

$$\beta = \alpha(\alpha - x_1) \qquad \text{and} \qquad \mathbf{v} = (x_1 - \alpha, x_2, \ldots, x_n)^T$$

In both computations we can avoid loss of significant digits by choosing α to have the opposite sign of x_1.

18.

$$\begin{aligned} ULU &= \begin{pmatrix} 1 & \frac{\cos\theta - 1}{\sin\theta} \\ 0 & 1 \end{pmatrix} \begin{pmatrix} 1 & 0 \\ \sin\theta & 1 \end{pmatrix} \begin{pmatrix} 1 & \frac{\cos\theta - 1}{\sin\theta} \\ 0 & 1 \end{pmatrix} \\ &= \begin{pmatrix} \cos\theta & \frac{\cos\theta - 1}{\sin\theta} \\ \sin\theta & 1 \end{pmatrix} \begin{pmatrix} 1 & \frac{\cos\theta - 1}{\sin\theta} \\ 0 & 1 \end{pmatrix} \\ &= \begin{pmatrix} \cos\theta & -\sin\theta \\ \sin\theta & \cos\theta \end{pmatrix} \end{aligned}$$

SECTION 6

3. (a) $\mathbf{v}_1 = A\mathbf{u}_0 = \begin{pmatrix} 3 \\ -2 \end{pmatrix} \qquad \mathbf{u}_1 = \frac{1}{3}\mathbf{v}_1 = \begin{pmatrix} 1 \\ -2/3 \end{pmatrix}$

$$\mathbf{v}_2 = A\mathbf{u}_1 = \begin{pmatrix} -1/3 \\ -1/3 \end{pmatrix} \qquad \mathbf{u}_2 = -3\mathbf{v}_2 = \begin{pmatrix} 1 \\ 1 \end{pmatrix}$$

$$\mathbf{v}_3 = A\mathbf{u}_2 = \begin{pmatrix} 3 \\ -2 \end{pmatrix} \qquad \mathbf{u}_3 = \frac{1}{3}\mathbf{v}_3 = \begin{pmatrix} 1 \\ -2/3 \end{pmatrix}$$

$$\mathbf{v}_4 = A\mathbf{u}_3 = \begin{pmatrix} -1/3 \\ -1/3 \end{pmatrix} \qquad \mathbf{u}_4 = -3\mathbf{v}_4 = \begin{pmatrix} 1 \\ 1 \end{pmatrix}$$

6. (a and b). Let $\mathbf{x}_j$ be an eigenvector of A belonging to λ_j.

$$B^{-1}\mathbf{x}_j = (A - \lambda I)\mathbf{x}_j = (\lambda_j - \lambda)\mathbf{x}_j = \frac{1}{\mu_j}\mathbf{x}_j$$

Multiplying through by $\mu_j B$ we obtain

$$B\mathbf{x}_j = \mu_j \mathbf{x}_j$$

Thus μ_j is an eigenvalue of B and $\mathbf{x}_j$ is an eigenvector belonging to μ_j.

(c) If λ_k is the eigenvalue of A that is closest to λ, then

$$|\mu_k| = \frac{1}{|\lambda_k - \lambda|} > \frac{1}{|\lambda_j - \lambda|} = |\mu_j|$$

for $j \neq k$. Therefore μ_k is the dominant eigenvalue of B. Thus when the power method is applied to B, it will converge to an eigenvector $\mathbf{x}_k$ of μ_k. By part (b), $\mathbf{x}_k$ will also be an eigenvector belonging to λ_k.

7. (a) Since $A\mathbf{x} = \lambda\mathbf{x}$, the ith coordinate of each side must be equal. Thus

$$\sum_{j=1}^{n} a_{ij}x_j = \lambda x_i$$

(b) It follows from part (a) that

$$(\lambda - a_{ii})x_i = \sum_{\substack{j=1 \\ j\neq i}}^{n} a_{ij}x_j$$

Since $|x_1| = \|\mathbf{x}\|_\infty > 0$ it follows that

$$|\lambda - a_{ii}| = \left| \sum_{\substack{j=1 \\ j\neq i}}^{n} \frac{a_{ij}x_j}{x_i} \right| \leq \sum_{\substack{j=1 \\ j\neq i}}^{n} |a_{ij}| \left| \frac{x_j}{x_i} \right| \leq \sum_{\substack{j=1 \\ j\neq i}}^{n} |a_{ij}|$$

8. (a) Let $B = X^{-1}(A+E)X$. Since $X^{-1}AX$ is a diagonal matrix whose diagonal entries are the eigenvalues of A we have

$$b_{ij} = \begin{cases} c_{ij} & \text{if } i \neq j \\ \lambda_i + c_{ii} & \text{if } i = j \end{cases}$$

It follows from Exercise 7 that

$$|\lambda - b_{ii}| \leq \sum_{\substack{j=1 \\ j\neq i}}^{n} |b_{ij}|$$

for some i. Thus

$$|\lambda - \lambda_i - c_{ii}| \leq \sum_{\substack{j=1 \\ j\neq i}}^{n} |c_{ij}|$$

Since

$$|\lambda - \lambda_i| - |c_{ii}| \leq |\lambda - \lambda_i - c_{ii}|$$

it follows that

$$|\lambda - \lambda_i| \leq \sum_{j=1}^{n} |c_{ij}|$$

(b) It follows from part (a) that

$$\begin{aligned}
\min_{1\le j\le n} |\lambda - \lambda_j| &\le \max_{1\le i\le n}\left(\sum_{j=1}^{n}|c_{ij}|\right)\\
&= \|C\|_\infty\\
&\le \|X^{-1}\|_\infty\|E\|_\infty\|X\|_\infty\\
&= \text{cond}_\infty(X)\|E\|_\infty
\end{aligned}$$

9. The proof is by induction on k. In the case $k = 1$

$$AP_1 = (Q_1R_1)Q_1 = Q_1(R_1Q_1) = P_1A_2$$

Assuming $P_mA_{m+1} = AP_m$ we will show that $P_{m+1}A_{m+2} = AP_{m+1}$.

$$\begin{aligned}
AP_{m+1} &= AP_mQ_{m+1}\\
&= P_mA_{m+1}Q_{m+1}\\
&= P_mQ_{m+1}R_{m+1}Q_{m+1}\\
&= P_{m+1}A_{m+2}
\end{aligned}$$

10. (a) The proof is by induction on k. In the case $k = 1$

$$P_2U_2 = Q_1Q_2R_2R_1 = Q_1A_2R_1 = P_1A_2U_1$$

It follows from Exercise 9 that

$$P_1A_2U_1 = AP_1U_1$$

Thus

$$P_2U_2 = P_1A_2U_1 = AP_1U_1$$

If

$$P_{m+1}U_{m+1} = P_mA_{m+1}U_m = AP_mU_m$$

then

$$\begin{aligned}
P_{m+2}U_{m+2} &= P_{m+1}Q_{m+2}R_{m+2}U_{m+1}\\
&= P_{m+1}A_{m+2}U_{m+1}
\end{aligned}$$

Again by Exercise 9 we have

$$P_{m+1}A_{m+2} = AP_{m+1}$$

Thus

$$P_{m+2}U_{m+2} = P_{m+1}A_{m+2}U_{m+1} = AP_{m+1}U_{m+1}$$

(b) Prove: $P_kU_k = A^k$. The proof is by induction on k. In the case $k = 1$

$$P_1U_1 = Q_1R_1 = A = A^1$$

If

$$P_mU_m = A^m$$

then it follows from part (a) that

$$P_{m+1}U_{m+1} = AP_mU_m = AA^m = A^{m+1}$$

11. To determine $\mathbf{x}_k$ and β, compare entries on both sides of the block multiplication for the equation $R_{k+1}U_{k+1} = U_{k+1}D_{k+1}$.

$$\begin{pmatrix} R_k & \mathbf{b}_k \\ \mathbf{0}^T & \beta_k \end{pmatrix} \begin{pmatrix} U_k & \mathbf{x}_k \\ \mathbf{0}^T & 1 \end{pmatrix} = \begin{pmatrix} U_k & \mathbf{x}_k \\ \mathbf{0}^T & 1 \end{pmatrix} \begin{pmatrix} D_k & \mathbf{0} \\ \mathbf{0}^T & \beta \end{pmatrix}$$

$$\begin{pmatrix} R_kU_k & R_k\mathbf{x}_k + b_k \\ \mathbf{0}^T & \beta_k \end{pmatrix} = \begin{pmatrix} U_kD_k & \beta\mathbf{x}_k \\ \mathbf{0}^T & \beta \end{pmatrix}$$

By hypothesis, $R_kU_k = U_kD_k$, so if we set $\beta = \beta_k$, then the diagonal blocks of both sides will match up. Equating the $(1,2)$ blocks of both sides we get

$$R_k\mathbf{x}_k + \mathbf{b}_k = \beta_k\mathbf{x}_k$$

$$(R_k - \beta_kI)\mathbf{x}_k = -\mathbf{b}_k$$

This is a $k \times k$ upper triangular system. The system has a unique solution since β_k is not an eigenvalue of R_k. The solution $\mathbf{x}_k$ can be determined by back substitution.

12. (a) Algorithm for computing eigenvectors of an $n \times n$ upper triangular matrix with no multiple eigenvalues.

Set $U_1 = (1)$
For $k = 1, \ldots, n-1$
  Use back substitution to solve

$$(R_k - \beta_kI)\,\mathbf{x}_k = -\mathbf{b}_k$$

  where

$$\beta_k = r_{k+1,k+1} \quad \text{and} \quad \mathbf{b}_k = (r_{1,k+1}, r_{2,k+1}, \ldots, r_{k,k+1})^T$$

  Set

$$U_{k+1} = \begin{pmatrix} U_k & \mathbf{x}_k \\ \mathbf{0}^T & 1 \end{pmatrix}$$

End (For Loop)

The matrix U_n is upper triangular with 1's on the diagonal. Its column vectors are the eigenvectors of R.

(b) All of the arithmetic is done in solving the $n-1$ systems

$$(R_k - \beta_kI)\mathbf{x}_k = -\mathbf{b}_k \qquad k = 1, \ldots, n-1$$

by back substitution. Solving the kth system requires

$$1 + 2 + \cdots + k = \frac{k(k+1)}{2} \text{ multiplications}$$

and k divisions. Thus the kth step of the loop requires $\frac{1}{2}k^2 + \frac{3}{2}k$ multiplications/divisions. The total algorithm requires

$$\frac{1}{2}\sum_{k=1}^{n-1}(k^2 + 3k) = \frac{1}{2}\left(\frac{n(2n-1)(n-1)}{6} + \frac{3n(n-1)}{2}\right)$$

$$= \frac{n^3}{6} + \frac{4n^2 - n - 4}{6} \text{ multiplications/divisions}$$

The dominant term is $n^3/6$.

SECTION 7

3. (a) $\alpha_1 = \|\mathbf{a}_1\| = 2,\ \beta_1 = \alpha_1(\alpha_1 - \alpha_{11}) = 2,\quad \mathbf{v}_1 = (-1,\ 1,\ 1,\ 1)^T$

$H_1 = I - \frac{1}{\beta_1}\mathbf{v}_1\mathbf{v}_1^T$

$$H_1 A = \begin{pmatrix} 2 & 3 \\ 0 & 2 \\ 0 & 1 \\ 0 & -2 \end{pmatrix} \qquad H_1\mathbf{b} = \begin{pmatrix} 8 \\ -1 \\ -8 \\ -5 \end{pmatrix}$$

$\alpha_2 = \|(2,\ 1,\ -2)^T\| = 3 \quad \beta_2 = 3(3-2) = 3 \quad \mathbf{v}_2 = (-1,\ 1,\ -2)^T$

$$H_2 = \begin{pmatrix} 1 & \mathbf{0}^T \\ \mathbf{0} & H_{22} \end{pmatrix} \text{ where } H_{22} = I - \frac{1}{\beta_2}\mathbf{v}_2\mathbf{v}_2^T$$

$$H_2H_1A = \begin{pmatrix} 2 & 3 \\ 0 & 3 \\ 0 & 0 \\ 0 & 0 \end{pmatrix} \qquad H_2H_1\mathbf{b} = \begin{pmatrix} 8 \\ 0 \\ -9 \\ -3 \end{pmatrix}$$

5. Let A be an $m \times n$ matrix with nonzero singular values $\sigma_1, \ldots, \sigma_r$ and singular value decomposition $U\Sigma V^T$. We will show first that Σ^+ satisfies the four Penrose conditions. Note that the matrix $\Sigma\Sigma^+$ is an $m \times m$ diagonal matrix whose first r diagonal entries are all 1 and whose remaining diagonal entries are all 0. Since the only nonzero entries in the matrices Σ and Σ^+ occur in the first r diagonal positions it follows that

$$(\Sigma\Sigma^+)\Sigma = \Sigma \quad\text{and}\quad \Sigma^+(\Sigma\Sigma^+) = \Sigma^+$$

Thus Σ^+ satisfies the first two Penrose conditions. Since both $\Sigma\Sigma^+$ and $\Sigma^+\Sigma$ are square diagonal matrices they must be symmetric

$$\begin{aligned} (\Sigma\Sigma^+)^T &= \Sigma\Sigma^+ \\ (\Sigma^+\Sigma)^T &= \Sigma^+\Sigma \end{aligned}$$

Thus Σ^+ satisfies all four Penrose conditions. Using this result it is easy to show that $A^+ = V\Sigma^+U^T$ satisfies the four Penrose conditions.

(1) $AA^+A = U\Sigma V^TV\Sigma^+U^TU\Sigma V^T = U\Sigma\Sigma^+\Sigma V^T = U\Sigma V^T = A$

(2) $A^+AA^+ = V\Sigma^+U^TU\Sigma V^TV\Sigma^+U^T = V\Sigma^+\Sigma\Sigma^+U^T = V\Sigma^+U^T = A^+$

(3)
$$\begin{aligned} (AA^+)^T &= (U\Sigma V^TV\Sigma^+U^T)^T \\ &= (U\Sigma\Sigma^+U^T)^T \\ &= U(\Sigma\Sigma^+)^TU^T \end{aligned}$$

$$\begin{aligned} &= U(\Sigma\Sigma^+)U^T \\ &= AA^+ \end{aligned}$$

(4) $$\begin{aligned} (A^+A)^T &= (V\Sigma^+U^TU\Sigma V^T)^T \\ &= (V\Sigma^+\Sigma V^T)^T \\ &= V(\Sigma^+\Sigma)^TV^T \\ &= V(\Sigma^+\Sigma)V^T \\ &= A^+A \end{aligned}$$

6. Let B be a matrix satisfying Penrose condition (1) and (3), that is,

$$ABA = A \qquad \text{and} \qquad (AB)^T = AB$$

If $\mathbf{x} = B\mathbf{b}$, then

$$A^TA\mathbf{x} = A^TAB\mathbf{b} = A^T(AB)^T\mathbf{b} = (ABA)^T\mathbf{b} = A^T\mathbf{b}$$

7. If $X = \frac{1}{\|\mathbf{x}\|_2^2}\mathbf{x}^T$, then

$$X\mathbf{x} = \frac{1}{\|\mathbf{x}\|_2^2}\mathbf{x}^T\mathbf{x} = 1$$

Using this it is easy to verify that $\mathbf{x}$ and X satisfy the four Penrose conditions.

(1) $\mathbf{x}X\mathbf{x} = \mathbf{x}1 = \mathbf{x}$

(2) $X\mathbf{x}X = 1X = X$

(3) $(\mathbf{x}X)^T = X^T\mathbf{x}^T = \frac{1}{\|\mathbf{x}\|^2}\mathbf{x}\mathbf{x}^T = \mathbf{x}X$

(4) $(X\mathbf{x})^T = 1^T = 1 = X\mathbf{x}$

8. If A has singular value decomposition $U\Sigma V^T$ then

(7) $$A^TA = V\Sigma^TU^TU\Sigma V^T = V\Sigma^T\Sigma V^T$$

The matrix $\Sigma^T\Sigma$ is an $n \times n$ diagonal matrix with diagonal entries $\sigma_1^2, \ldots, \sigma_n^2$. Since A has rank n its singular values are all nonzero and it follows that $\Sigma^T\Sigma$ is nonsingular. It follows from equation (7) that

$$\begin{aligned} (A^TA)^{-1}A^T &= (V(\Sigma^T\Sigma)^{-1}V^T)(V\Sigma^TU^T) \\ &= V(\Sigma^T\Sigma)^{-1}\Sigma^TU^T \\ &= V\Sigma^+U^T \\ &= A^+ \end{aligned}$$

9. Let

$$\mathbf{b} = AA^+\mathbf{b} = A(A^+\mathbf{b})$$

since

$$R(A) = \{A\mathbf{x} \mid \mathbf{x} \in R^n\}$$

it follows that $\mathbf{b} \in R(A)$.

Conversely if $\mathbf{b} \in R(A)$, then $\mathbf{b} = A\mathbf{x}$ for some $\mathbf{x} \in R^n$. It follows that

$$\begin{aligned} A^+\mathbf{b} &= A^+A\mathbf{x} \\ AA^+\mathbf{b} &= AA^+A\mathbf{x} = A\mathbf{x} = \mathbf{b} \end{aligned}$$

10. A vector $\mathbf{x} \in R^n$ minimizes $\|\mathbf{b} - A\mathbf{x}\|_2$ if and only if $\mathbf{x}$ is a solution to the normal equations. It follows from Theorem 7.9.1 that $A^+\mathbf{b}$ is a particular solution. Since $A^+\mathbf{b}$ is a particular solution it follows that a vector $\mathbf{x}$ will be a solution if and only if

$$\mathbf{x} = A^+\mathbf{b} + \mathbf{z}$$

where $\mathbf{z} \in N(A^TA)$. However, $N(A^TA) = N(A)$. Since $\mathbf{v}_{r+1}, \ldots, \mathbf{v}_n$ form a basis for $N(A)$ (see Exercise 7, Section 7), it follows that $\mathbf{x}$ is a solution if and only if

$$\mathbf{x} = A^+\mathbf{b} + c_{r+1}\mathbf{v}_{r+1} + \cdots + c_n\mathbf{v}_n$$

13. (a) $(\Sigma^+)^+$ is an $m \times n$ matrix whose nonzero diagonal entries are the reciprocals of the nonzero diagonal entries of Σ^+. Thus $(\Sigma^+)^+ = \Sigma$. If $A = U\Sigma V^T$, then

$$(A^+)^+ = (V\Sigma^+U^T)^+ = U(\Sigma^+)^+V^T = U\Sigma V^T = A$$

(b) $\Sigma\Sigma^+$ is an $m \times m$ diagonal matrix whose diagonal entries are all 0's and 1's. Thus $(\Sigma\Sigma^+)^2 = \Sigma\Sigma^+$ and it follows that

$$\begin{aligned} (AA^+)^2 &= (U\Sigma V^TV\Sigma^+U^T)^2 = (U\Sigma\Sigma^+U^T)^2 = U(\Sigma\Sigma^+)^2U^T \\ &= U\Sigma\Sigma^+U^T = AA^+ \end{aligned}$$

(c) $\Sigma^+\Sigma$ is an $n \times n$ diagonal matrix whose diagonal entries are all 0's and 1's. Thus $(\Sigma^+\Sigma)^2 = \Sigma^+\Sigma$ and it follows that

$$\begin{aligned} (A^+A)^2 &= (V\Sigma^+U^TU\Sigma V^T)^2 = (V\Sigma^+\Sigma V^T)^2 = V(\Sigma^+\Sigma)^2V^T \\ &= V\Sigma^+\Sigma V^T = A^+A \end{aligned}$$

15. (1)
$$\begin{aligned} ABA &= XY^T[Y(Y^TY)^{-1}(X^TX)^{-1}X^T]XY^T \\ &= X(Y^TY)(Y^TY)^{-1}(X^TX)^{-1}(X^TX)Y^T \\ &= XY^T \\ &= A \end{aligned}$$

(2)
$$\begin{aligned} BAB &= [Y(Y^TY)^{-1}(X^TX)^{-1}X^T](XY^T)[Y(Y^TY)^{-1}(X^TX)^{-1}X^T] \\ &= Y(Y^TY)^{-1}(X^TX)^{-1}(X^TX)(Y^TY)(Y^TY)^{-1}(X^TX)^{-1}X^T \\ &= Y(Y^TY)^{-1}(X^TX)^{-1}X^T \\ &= B \end{aligned}$$

(3)
$$\begin{aligned} (AB)^T &= B^TA^T \\ &= [Y(Y^TY)^{-1}(X^TX)^{-1}X^t]^T(YX^T) \\ &= X(X^TX)^{-1}(Y^TY)^{-1}Y^TYX^T \\ &= X(X^TX)^{-1}X^T \\ &= X(Y^TY)(Y^TY)^{-1}(X^TX)^{-1}X^T \end{aligned}$$

$$
\begin{aligned}
&= (XY^T)[Y(Y^TY)^{-1}(X^TX)^{-1}X^T] \\
&= AB \\
\text{(4) } (BA)^T &= A^TB^T \\
&= (YX^T)[Y(Y^TY)^{-1}(X^TX)^{-2}X^T]^T \\
&= YX^TX(X^TX)^{-1}(Y^TY)^{-1}Y^T \\
&= Y(Y^TY)^{-1}Y^T \\
&= Y(Y^TY)^{-1}(X^TX)^{-1}(X^TX)Y^T \\
&= [Y(Y^TY)^{-1}(X^TX)^{-1}X^T](XY^T) \\
&= BA
\end{aligned}
$$

MATLAB EXERCISES

1. The system is well conditioned since perturbations in the solutions are roughly the same size as the perturbations in A and $\mathbf{b}$.

2. (a) The entries of $\mathbf{b}$ and the entries of $V\mathbf{s}$ should both be equal to the row sums of V.

3. (a) Since L is lower triangular with 1's on the diagonal, it follows that $\det(L) = 1$ and

$$\det(C) = \det(L)\det(L^T) = 1$$

and hence $C^{-1} = \text{adj}(C)$. Since C is an integer matrix its adjoint will also consist entirely of integers.

7. Since A is a magic square, the row sums of $A - tI$ will all be 0. Thus the row vectors of $A - tI$ must be linearly dependent. Therefore $A - tI$ is singular and hence t is an eigenvalue of A. Since the sum of all the eigenvalues is equal to the trace, the other eigenvalues must add up to 0. The condition number of X should be small, which indicates that the eigenvalue problem is well-conditioned.

8. Since A is upper triangular no computations are necessary to determine its eigenvalues. Thus MATLAB will give you the exact eigenvalues of A. However the eigenvalue problem is moderately ill-conditioned and consequently the eigenvalues of A and $A1$ will differ substantially.

9. (b) Cond(X) should be on the order of 10^8, so the eigenvalue problem should be moderately ill-conditioned.

10. (b) $K\mathbf{e} = -H\mathbf{e}$.

12. (a) The graph has been rotated 45° in the counterclockwise direction.

(c) The graph should be the same as the graph from part (b). Reflecting about a line through the origin at an angle of $\frac{\pi}{8}$ is geometrically the same as reflecting about the x-axis and then rotating 45 degrees. The later pair of operations can be represented by the matrix product

$$\begin{pmatrix} c & -s \\ s & c \end{pmatrix}\begin{pmatrix} 1 & 0 \\ 0 & -1 \end{pmatrix} = \begin{pmatrix} c & s \\ s & -c \end{pmatrix}$$

where $c = \cos\frac{\pi}{4}$ and $s = \sin\frac{\pi}{4}$.

13. (b)

$$\mathbf{b}(1,:) = \mathbf{b}(2,:) = \mathbf{b}(3,:) = \mathbf{b}(4,:) = \tfrac{1}{2}(\mathbf{a}(2,:) + \mathbf{a}(3,:))$$

(c) Both A and B have the same largest singular value $s(1)$. Therefore

$$\|A\|_2 = s(1) = \|B\|_2$$

The matrix B is rank 1. Therefore $s(2) = s(3) = s(4) = 0$ and hence

$$\|B\|_F = \|\mathbf{s}\|_2 = s(1)$$

14. (b)

$$\|A\|_2 = s(1) = \|B\|_2$$

(c) To construct C, set

$$D(4,4) = 0 \qquad \text{and} \qquad C = U * D * V'$$

It follows that

$$\|C\|_2 = s(1) = \|A\|_2$$

and

$$\|C\|_F = \sqrt{s(1)^2 + s(2)^2 + s(3)^2} < \|\mathbf{s}\|_2 = \|A\|_F$$

15. (a) The rank of A should be 4. To determine $V1$ and $V2$ set

$$V1 = V(:, 1:4) \qquad V2 = V(:, 5:6)$$

P is the projection matrix onto $N(A)$. Therefore $\mathbf{r}$ must be in $N(A)$. Since $\mathbf{w} \in R(A^T) = N(A)^\perp$, we have

$$\mathbf{r}^T\mathbf{w} = 0$$

(b) Q is the projection matrix onto $N(A^T)$. Therefore $\mathbf{y}$ must be in $N(A^T)$. Since $\mathbf{z} \in R(A) = N(A^T)^\perp$, we have

$$\mathbf{y}^T\mathbf{z} = 0$$

(d) Both AX and $U1(U1)^T$ are projection matrices onto $R(A)$. Since the projection matrix onto a subspace is unique, it follows that

$$AX = U1(U1)^T$$

16. (b) The disk centered at 50 is disjoint from the other two disks, so it contains exactly one eigenvalue. The eigenvalue is real so it must lie in the interval $[46, 54]$. The matrix C is similar to B and hence must have the same eigenvalues. The disks of C centered at 3 and 7 are disjoint from the other disks. Therefore each of the two disks contains an eigenvalue. These eigenvalues are real and consequently must lie in the intervals $[2.7, 3.3]$ and $[6.7, 7.3]$. The matrix C^T has the same eigenvalues as C and B. Using the Gerschgorin disk corresponding to the third row of C^T we see that the dominant eigenvalue must lie in the interval $[49.6, 50.4]$. Thus without computing the eigenvalues of B we are able to obtain nice approximations to their actual locations.

CHAPTER TEST A

1. The statement is false in general. For example, if
$$a = 0.11 \times 10^0, \quad b = 0.32 \times 10^{-2}, \quad c = 0.33 \times 10^{-2}$$
and 2-digit decimal arithmetic is used, then
$$fl(fl(a+b)+c) = a = 0.11 \times 10^0$$
and
$$fl(a + fl(b+c)) = 0.12 \times 10^0$$
2. The statement is false in general. For example, if A and B are both 2×2 matrices and C is a 2×1 matrix, then the computation of $A(BC)$ requires 8 multiplications and 4 additions, while the computation of $(AB)C$ requires 12 multiplications and 6 additions.
3. The statement is false in general. It is possible to have a large relative error if the coefficient matrix is ill-conditioned. For example, the $n \times n$ Hilbert matrix H is defined by
$$h_{ij} = \frac{1}{i+j-1}$$
For $n = 12$, the matrix H is nonsingular, but it is very ill-conditioned. If you tried to solve a nonhomogeneous linear system with this coefficient matrix you would not get an accurate solution.
4. The statement is true. For a symmetric matrix the eigenvalue problem is well conditioned. (See the remarks following Theorem 7.6.1.) If a stable algorithm is used then the computed eigenvalues should be the exact eigenvalues of a nearby matrix, i.e., a matrix of the form $A + E$ where $\|E\|$ is small. Since the problem is well conditioned the eigenvalues of nearby matrices will be good approximations to the eigenvalues of A.
5. The statement is false in general. If the matrix is nonsymmetric then the eigenvalue problem could be ill-conditioned. If so, then even a stable algorithm will not necessary guarantee accurate eigenvalues. In particular if A has an eigenvalue-eigenvector decomposition XDX^{-1} and X is very ill-conditioned, then the eigenvalue problem will be ill-conditioned and it will not be possible to compute the eigenvalues accurately.
6. The statement is false. If A^{-1} and the LU factorization are both available the it doesn't matter which you use since it takes the same number of arithmetic operations to solve $LU\mathbf{x} = \mathbf{b}$ using forward and back substitution as it does to multiply $A^{-1}\mathbf{b}$.
7. The statement is true. The 1-norm is computed by taking the sum of the absolute values on the entries in each column of A and then taking the maximum of the column sums. The infinity norm is computed by taking the sum of the absolute values on the entries in each row of A and then taking the maximum of the row sums. If A is symmetric then the row sums and column sums will be the same and hence the both norms will be equal.

8. The statement is false in general. For example if

$$A = \begin{pmatrix} 4 & 0 \\ 0 & 3 \end{pmatrix}$$

then $\|A\|_2 = 4$ and $\|A\|_F = 5$.

9. The statement is false in general. If A has rank n, then the least squares problem will have a unique solution. However, if A is ill-conditioned the computed solution may not be a good approximation to the exact solution even though it produces a small residual vector.

10. The statement is false in general. For example, if

$$A = \begin{pmatrix} 1 & 0 \\ 0 & 10^{-8} \end{pmatrix} \quad \text{and} \quad B = \begin{pmatrix} 1 & 0 \\ 0 & 0 \end{pmatrix}$$

then A and B are close since $\|A - B\|_F = 10^{-8}$. However their pseudoinverses are not close. In fact, $\|A^+ - B^+\|_F = 10^8$

CHAPTER TEST B

1. If $\mathbf{y} = B\mathbf{x}$ then the computation of a single entry of $\mathbf{y}$ requires n multiplications and $n-1$ additions. Since $\mathbf{y}$ has n entries, the computation of the matrix-vector product $B\mathbf{x}$ requires n^2 multiplications and $n(n-1)$ additions. The computation $A(B\mathbf{x}) = A\mathbf{y}$ requires 2 matrix-vector multiplications. So the number of scalar multiplications and scalar additions that are necessary is $2n^2$ and $2n(n-1)$.

On the other hand if $C = AB$ then the computation of the jth column of C requires a matrix-vector multiplication $\mathbf{c}_j = A\mathbf{b}_j$ and hence the computation of C requires n matrix-vector multiplications. Therefore the computation $(AB)\mathbf{x} = C\mathbf{x}$ will require $n+1$ matrix-vector multiplications. The total number of arithmetic operations will be $(n+1)n^2$ scalar multiplications and $(n+1)n(n-1)$ scalar additions.

For $n > 1$ the computation $A(B\mathbf{x})$ is more efficient.

2. **(a)**

$$\left(\begin{array}{ccc|c} 2 & 3 & 6 & 3 \\ 4 & 4 & 8 & 0 \\ 1 & 3 & 4 & 4 \end{array}\right) \rightarrow \left(\begin{array}{ccc|c} 4 & 4 & 8 & 0 \\ 2 & 3 & 6 & 3 \\ 1 & 3 & 4 & 4 \end{array}\right) \rightarrow \left(\begin{array}{ccc|c} 4 & 4 & 8 & 0 \\ 0 & 1 & 2 & 3 \\ 0 & 2 & 2 & 4 \end{array}\right)$$

$$\rightarrow \left(\begin{array}{ccc|c} 4 & 4 & 8 & 0 \\ 0 & 2 & 2 & 4 \\ 0 & 1 & 2 & 3 \end{array}\right) \rightarrow \left(\begin{array}{ccc|c} 4 & 4 & 8 & 0 \\ 0 & 2 & 2 & 4 \\ 0 & 0 & 1 & 1 \end{array}\right)$$

The solution $\mathbf{x} = (-3, 1, 1)^T$ is obtained using back substitution.

(b)

$$P = \begin{pmatrix} 0 & 1 & 0 \\ 0 & 0 & 1 \\ 1 & 0 & 0 \end{pmatrix}, \quad PA = \begin{pmatrix} 4 & 4 & 8 \\ 1 & 3 & 4 \\ 2 & 3 & 6 \end{pmatrix}$$

and

$$LU = \begin{pmatrix} 1 & 0 & 0 \\ \frac{1}{4} & 1 & 0 \\ \frac{1}{2} & \frac{1}{2} & 1 \end{pmatrix} \begin{pmatrix} 4 & 4 & 8 \\ 0 & 2 & 2 \\ 0 & 0 & 1 \end{pmatrix}$$

(c) If we set $\mathbf{d} = P\mathbf{c} = (8,2,1)^T$ and solve $L\mathbf{y} = \mathbf{d}$ by forward substitution

$$(L \;\; \mathbf{d}) = \left(\begin{array}{ccc|c} 1 & 0 & 0 & 8 \\ \frac{1}{4} & 1 & 0 & 2 \\ \frac{1}{2} & \frac{1}{2} & 1 & 1 \end{array}\right)$$

then the solution is $\mathbf{y} = (8, 0, -3)^T$. To find the solution to the system $A\mathbf{x} = \mathbf{c}$, we solve $U\mathbf{x} = \mathbf{y}$ using back substitution.

$$(U \;\; \mathbf{y}) = \left(\begin{array}{ccc|c} 4 & 4 & 8 & 8 \\ 0 & 2 & 2 & 0 \\ 0 & 0 & 1 & -3 \end{array}\right)$$

The solution is $\mathbf{x} = (5, 3, -3)^T$.

3. If Q is a 4×4 orthogonal matrix then for any nonzero $\mathbf{x}$ in R^4 we have $\|Q\mathbf{x}\| = \|\mathbf{x}\|$ and hence

$$\|Q\|_2 = \max_{\mathbf{x} \neq \mathbf{0}} \frac{\|Q\mathbf{x}\|}{\|\mathbf{x}\|} = 1$$

To determine the Frobenius norm of Q, note that

$$\|Q\|_F^2 = \|\mathbf{q}_1\|^2 + \|\mathbf{q}_2\|^2 + \|\mathbf{q}_3\|^2 + \|\mathbf{q}_4\|^2 = 4$$

and hence $\|Q\|_F = 2$.

4. (a) $\|H\|_1 = 1 + \frac{1}{2} + \frac{1}{3} + \frac{1}{4} = \frac{25}{12}$

$\|H^{-1}\|_1 = \max(516,\ 5700,\ 13620,\ 8820) = 13620$

(b) From part (a) we have $\text{cond}_1(H) = \frac{25}{12} \cdot 13620 = 28375$ and hence

$$\frac{\|\mathbf{x} - \mathbf{x}'\|_1}{\|\mathbf{x}\|_1} \leq \text{cond}_1(H) \frac{\|\mathbf{r}\|}{\|\mathbf{b}\|_1} = 28375 \cdot \frac{0.36 \times 10^{-11}}{50} = 2.043 \times 10^{-9}$$

5. The relative error in the solution is bounded by

$$\text{cond}_\infty(A) \frac{\|\mathbf{r}\|_\infty}{\|\mathbf{b}\|_\infty} \approx 10^7 \epsilon$$

so it is possible that one could lose as many as 7 digits of accuracy.

6. **(a)** $\alpha = 3$, $\beta = 3(3-1) = 6$, $\mathbf{v} = (-2, 2, -2)^T$

$$H = I - \frac{1}{\beta}\mathbf{v}\mathbf{v}^T = \begin{pmatrix} \frac{1}{3} & \frac{2}{3} & -\frac{2}{3} \\ \frac{2}{3} & \frac{1}{3} & \frac{2}{3} \\ -\frac{2}{3} & \frac{2}{3} & \frac{1}{3} \end{pmatrix}$$

(b)

$$G = \begin{pmatrix} 1 & 0 & 0 \\ 0 & \frac{1}{\sqrt{2}} & -\frac{1}{\sqrt{2}} \\ 0 & -\frac{1}{\sqrt{2}} & -\frac{1}{\sqrt{2}} \end{pmatrix}$$

7. If A has QR-factorization $A = QR$ and $B = RQ$ then

$$Q^TAQ = Q^TQRQ = RQ = B$$

The matrices A and B are similar and consequently must have the same eigenvalues. Furthermore, if λ is an eigenvalue of B and $\mathbf{x}$ is an eigenvector belonging to λ then

$$Q^TAQ\mathbf{x} = B\mathbf{x} = \lambda\mathbf{x}$$

and hence

$$AQ\mathbf{x} = \lambda Q\mathbf{x}$$

So $Q\mathbf{x}$ is an eigenvector of A belonging to λ.

8. The estimate you get will depend upon your choice of a starting vector. If we start with $\mathbf{u}_0 = \mathbf{x}_0 = \mathbf{e}_1$, then

$$\begin{aligned}
&\mathbf{v}_1 = A\mathbf{e}_1 = \mathbf{a}_1, && \mathbf{u}_1 = \tfrac{1}{4}\mathbf{v}_1 = (0.25, 1)^T \\
&\mathbf{v}_2 = A\mathbf{u}_1 = (2.25, 4)^T, && \mathbf{u}_2 = \tfrac{1}{4}\mathbf{v}_2 = (0.5625, 1)^T \\
&\mathbf{v}_3 = A\mathbf{u}_2 = (2.5625, 5.25)^T, && \mathbf{u}_3 = \tfrac{1}{5.25}\mathbf{v}_3 = (0.548810, 1)^T \\
&\mathbf{v}_4 = A\mathbf{u}_3 = (2.48810, 4.95238)^T, && \mathbf{u}_4 = (0.502404, 1)^T \\
&\mathbf{v}_5 = A\mathbf{u}_4 = (2.50240, 5.00962)^T, && \mathbf{u}_5 == (0.499520, 1)^T \\
&\mathbf{v}_6 = A\mathbf{u}_5 = (2.49952, 4.99808)^T
\end{aligned}$$

Our computed eigenvalue is the second coordinate of $\mathbf{v}_6$, 4.99808 (rounded to 6 digits) and the computed eigenvector is $\mathbf{u}_5$. The actual dominant eigenvalue of A is $\lambda = 5$ and $\mathbf{x} = (0.5, 1)^T$ is an eigenvector belonging to λ.

9. The least squares solution with the smallest 2-norm is

$$\mathbf{x} = A^+\mathbf{b} = V\Sigma^+U^T\mathbf{b} = \begin{pmatrix} \frac{1}{3} \\ \frac{1}{2} \\ \frac{1}{12} \end{pmatrix}$$

10. If we set $\alpha_1 = \|\mathbf{a}_1\| = 2$, $\beta_1 = 2$, $\mathbf{v}_1 = (-1, 1, 1, 1)^T$ and $H_1 = I - \frac{1}{\beta_1}\mathbf{v}_1\mathbf{v}_1^T$ then $H_1\mathbf{a}_1 = 2\mathbf{e}_1$. If we multiply the augmented matrix $(A\ \ \mathbf{b})$ by H_1 we get

$$H_1(A\ |\mathbf{b}) = \left(\begin{array}{cc|c} 2 & 9 & 7 \\ 0 & 1 & -1 \\ 0 & 2 & 0 \\ 0 & -2 & -2 \end{array}\right)$$

Next we construct a 3×3 Householder matrix H_2 to zero out the last 2 entries of the vector $(1, 2, -2)^T$. If we set $\alpha_2 = 3$, $\beta_2 = 6$ and $\mathbf{v}_2 = (-2, 2, -2)^T$, then $H_2 = I - \frac{1}{\beta}\mathbf{v}_2\mathbf{v}_2^T$. If we apply H_2 to the last 3 rows of $H_1(A\ |\mathbf{b})$ we end up with the matrix

$$\left(\begin{array}{cc|c} 2 & 9 & 7 \\ 0 & 3 & 1 \\ 0 & 0 & -2 \\ 0 & 0 & 0 \end{array}\right)$$

The first two rows of this matrix form a triangular system. The solution $\mathbf{x} = (2, \frac{1}{3})^T$ to the triangular system is the solution to the least squares problem.